The Molecules of Life A Journey into Chemistry

Michael Johnson

Copyright © [2023]

Author: Michael Johnson

Title: The Molecules of Life A Journey into Chemistry

All rights reserved. No part of this book may be reproduced or transmitted in any form or by any means, electronic or mechanical, including photocopying, recording, or by any information storage and retrieval system, without permission in writing from the author.

This book is a product of

ISBN:

Table of content

Chapter name	Page No
1. Chemical and Molecular Basics:	1
2. Elements Essential to Life:	9
3. Water, the Life-Giver:	20
4. The Energy Molecules, Carbohydrates	32
5. Lipids, or "Life Fats,"	43
6. The proteins are the cellular labourers.	53
7. Nucleic Acids, the Carrier Particles of Information:	65
8. Chemical Reactions Within Cells, or Metabolism	75
9. Signalling molecules and hormones:	86
10. Health and Chemistry:	98
11. Chemistry's Bright Future in Life Sciences	110
12. Final Thoughts on an Ongoing Trip:	125

Chapter 1.
Chemical and Molecular Basics:

1.1 - An overview of chemistry's role in understanding life.

The Importance of Chemistry to Our Knowledge of Life

To bridge the gap between the microscopic world of atoms and molecules and the larger world we experience in everyday life, chemistry is sometimes referred to as the central science. It is the study of matter and all the ways in which it can and does change. Understanding the fundamentals of life is where chemistry's importance shines brightest.

Despite its seeming complexity, life is primarily a chemical phenomenon. All forms of life, from the tiniest single-celled creatures to the most complex human body, are driven by chemical reactions. Exploring the connection between chemistry and life is like making a pilgrimage to the centre of the universe because it is not coincidence.

Atoms, the fundamental particles of matter, are the basis of all biochemical processes. Hydrogen, along with carbon, oxygen, and nitrogen, and every other element on the periodic table, all contribute to the formation of the molecules upon which life depends. Each element's atomic structure determines some of its distinctive properties, which in turn affect the molecules and compounds it can create.

In turn, molecules are the life's molecular machinery. It is the unique arrangement and interactions of these atoms that give rise to the wide variety of life forms. For example, water, a molecule with only two hydrogen atoms and one oxygen atom (H2O), is crucial to life because of its amazing qualities, such as its solubility of a wide variety of chemicals and its ability to maintain a constant temperature. The biochemistry of living things is profoundly affected by these features.

The four major classes of biomolecules that control all of life's operations are carbohydrates, lipids, proteins, and nucleic acids. Carbohydrates, such glucose and cellulose, play crucial roles in cellular function and structure. Membranes surround and protect cells and organelles; they are composed of lipids like fats and phospholipids, which also allow for selective permeability.

However, proteins are the cells' main workers. Enzymes are incredibly adaptable molecules that can catalyse a wide variety of chemical reactions, as well as perform structural, transport, and signalling roles. Amino acids fold into three-dimensional structures to form proteins, which have a wide range of shapes and functions depending on their precise sequence.

DNA and RNA are two examples of nucleic acids that serve as information carriers in the living world. DNA holds the genetic information that define an organism's characteristics and traits, while RNA is essential in translating those instructions into the production of proteins. DNA's structure, a double helix made of nucleotide units, reveals the mechanisms behind heredity and evolution.

Perhaps most obviously, biochemistry is where chemists and biologists work together to study the molecular systems that control life. In order to understand how foods are metabolised into energy or building blocks for growth and repair, biochemists investigate the complex metabolic pathways that maintain life. Highly specialised proteins known as enzymes serve as catalysts in these reactions, reducing the activation energy needed for chemical activities to take place under the strict conditions of a living organism.

The study of chemistry is essential to unravelling the intricate network of intracellular signalling and communication. Complex chemical reactions yield hormones, neurotransmitters, and other signalling chemicals. Their release, reception, and subsequent molecular responses regulate different aspects of growth, development, and homeostasis, and are thus fundamental to important physiological activities.

There is no denying chemistry's importance in the field of medicine and healthcare. Pharmaceuticals are developed to treat diseases and reduce human suffering by zeroing down on specific molecules or pathways. New medications, improved drug formulations, and treatments for previously untreatable diseases are all the result of the joint efforts of chemists and biochemists.

Chemistry's influence extends far beyond the confines of the laboratory and the realm of medicine. The chemical makeup and responses of the things around us, including the food we eat, the clothes we wear, the water we drink, and the air we breathe, influence our experiences and environments. Knowing something about chemistry can help us make better choices regarding the things we buy, the food we eat, and the environmental effects of our actions.

Furthermore, chemistry cannot be divorced from ecological study. Understanding pollution, climate change, and the viability of ecosystems is aided by research into chemical reactions in the air, water, and soil. Monitoring environmental quality and devising measures to mitigate and avoid environmental degradation are both made possible by the analytical techniques made available by chemistry.

Understanding the complex interplay between chemistry and life is a never-ending quest. From the microscopic intricacy of cells to the planetary scale of ecological relationships, scientific research consistently reveals new levels of complexity. The study of chemistry is in a perpetual state of flux, expanding our horizons and enabling the development of cutting-edge tools and techniques.

In sum, the contributions of chemistry to our knowledge of the living world are immense and far-reaching. The study of molecular structure and function is crucial to understanding how life works and what it means. Any exploration into chemistry is a voyage into the very essence of existence, revealing insights into the complexity of biological processes, the creation of life-saving pharmaceuticals, and the maintenance of the health of our planet.

1.2- Introduction to the fundamental concept of molecules.

An Outline of the Basic Idea of Molecules

Molecules are the threads that connect the fabric of space and time, telling the story of matter and life. Everything we see, feel, and experience rests on the dance and interaction of these unseen creatures. Exploring the atomic level of thought is like removing layers of reality to get to the heart of chemistry.

Chemistry is the study of matter and the changes that it experiences, including its properties, content, and structure. It is a field that helps us make sense of the universe on both the macroscopic scale and the subatomic level. Along the way, we'll learn more about molecules — the smallest building blocks of matter — and how they play a crucial role in constructing our cosmos.

The fundamental entities of matter are atoms.

Understanding molecules requires a foundational knowledge of atoms, the fundamental building blocks of all things. Chemical language is written in atoms, which are the elemental characters of the cosmic alphabet.

The development of quantum physics in the early 20th century radically altered our perception of subatomic particles. We learned that atoms consist of a nucleus, which is positively charged and includes protons and neutrons, and an outermost cloud, which is negatively charged and contains a large number of electrons. These electrons are located in different "orbits," or energy levels, each of which has its own unique energy and geographical distribution.

The chemical characteristics of an element are determined by the way its electrons are arranged inside these energy levels. For instance, the identity of an element is determined by its atomic number, which is the total number of protons in its nucleus. Hydrogen has the lowest atomic number (1) because its nucleus contains only one proton. The atomic number of oxygen, an essential element in the air we breathe, is 8, indicating that it has eight protons in its nucleus.

Connecting atoms through chemical bonds.

The sharing or transfer of electrons between atoms is what makes chemistry so magical. Molecules are held together by these forces of attraction, called chemical bonds. To fully appreciate molecules, an appreciation of chemical bonds is required.

Covalent and ionic connections are the most frequent forms of chemical bonding. Molecules are formed when atoms share electrons, a process known as covalent bonding, and are then held together by the attractive force between the atoms. Covalent molecules, like water (H_2O), are extremely stable because two hydrogen atoms share electrons with one oxygen atom.

On the other hand, ions are formed by the process of electron transfer in ionic connections. Common table salt (NaCl) is an example of an ionic bond. The reaction between sodium (Na) and chlorine (Cl) results in the formation of two ions with opposite charges: a positively charged sodium ion (Na^+) and a negatively charged chloride ion (Cl^-), which are attracted to each other.

The molecules are the masterpieces of nature.

We can go more thoroughly into the concept of molecules now that we know how atoms interact with one another through chemical bonds. The products of these interactions are molecules, which are made up of atoms bonded to one another in a certain configuration. They range from elementary particles like oxygen (O_2) to complex multi-stranded polymers like DNA.

The vastness of the chemical cosmos is reflected in the incredible variety of molecules. Every molecule is special in its own way, with its own set of characteristics and capabilities. Think about the distinction between two molecules: one of water and one of methane (CH4). Water's surface tension and ability to act as a solvent are only two examples of how its atypical charge distribution gives rise to its distinctive features as a polar molecule. On the other hand, methane's nonpolarity influences its properties and interactions with other chemicals.

Moreover, molecules are not immobile objects but rather are in a continuous state of mobility. Molecules contain atoms that are free to move, rotate, and vibrate in relation to one another. The interactions between molecules and their surroundings can only be comprehended by gaining an appreciation for these dynamic motions.

What Makes Life Possible:

There is a staggering variety of compounds in the molecular world, yet a select few are of critical importance due to their roles in biology and the maintenance of life. Included in this category are the "molecules of life," which comprise carbohydrates, lipids, proteins, and nucleic acids.

Carbohydrates are carbon, hydrogen, and oxygen molecules that make up organic compounds. They are essential to the structural integrity of living things and provide their principal source of energy. In a process known as cellular respiration, cells convert glucose, a simple sugar, into energy.

Cell and organelle membranes are composed of lipids, sometimes known as fats, which are hydrophobic molecules. They can also serve as a form of energy storage. Lipids, and phospholipids in particular, play a crucial role in the creation of cell membranes, which are responsible for separating the inside of cells from their surroundings.

Amino acids are the building blocks of proteins, and peptide bonds are what hold them together. Enzymes, antibodies, and structural elements are only some of their many roles. Proteins' unique roles are established by the combination of their amino acid sequence and their three-dimensional structure.

Information is stored, passed on, and expressed with the help of nucleic acids like DNA (deoxyribonucleic acid) and RNA (ribonucleic acid). James Watson and Francis Crick's 1953 discovery of the DNA double helix was a watershed point in the development of molecular biology.

Finishing Up: The Nubbin of Chemistry

Molecules are the complicated patterns that arise from the interaction of atoms, and they play a crucial role in the fabric of existence. In them, chemistry expresses the mysteries of the cosmos to the world. Learning about molecules is like cracking the code of life, as they are the building blocks of all matter and the essence of chemistry. To truly understand the chemical cosmos, where every atom and link tells a story of creation, transformation, and connectivity, this voyage into the concept of molecules is only the beginning.

Chapter 2.
Elements Essential to Life:

2.1- Exploring the atoms and elements that compose molecules.

Investigating the Substances Made Up of Individual Atoms and Elements

To appreciate molecules and their far-reaching effects, we need to go back to the building blocks of matter: the atoms and elements. These atomic components are the fundamental units of chemistry, the raw materials from which all other compounds are made. Along the way, we'll explore the minute details of atoms and elements, learning about the amazing variety and cohesion that make up molecules.

The atom is the smallest unit in nature.

Atoms, the smallest units that maintain the properties of an element, are the fundamental building blocks of all matter. They make up everything we can see, feel, and experience; they are the chemistry's basic units. While atoms may only show up as tiny dots on scientific graphs, their complexity and beauty is truly astounding.
The three fundamental constituents of every atom are the proton, the neutron, and the electron. Electrons are negatively charged, neutrons have no charge, and protons are positively charged. The exact configuration of these particles within an atom's structure determines its chemical properties and uniqueness.

Protons and neutrons are located in the nucleus, a compact core in the atom's centre. The number of protons in the nucleus of an atom is what gives it its unique identity; this number is known as the atom's atomic number. Carbon, which has six protons, and oxygen, which has eight, are two common examples. However, neutrons only add to the atom's mass and not its uniqueness.

The electrons in a molecule's outermost layers follow different paths around the nucleus, each corresponding to a different energy level. The atom's interactions with others affect the way the electrons in its outer shells travel in their pairs and groups.

The Elements Are the Chemical Alphabet

Elements are the chemical language's equivalent of letters. Each element may be identified by its unique atomic number, which is determined by the number of protons in its nucleus. There are 118 known elements in the periodic table at present, and each one has its own unique properties and behaviours.

The elements have been graphically represented in a systematic fashion on the periodic table. Elements are classified into categories according to their shared qualities and properties. In the periodic table, the noble gases, which include helium and neon, may be found in the far right column. These elements are relatively unreactive and inert. In contrast, alkali metals like sodium and potassium react violently with water.

Elements are the universe's fundamental ingredients, and they may be combined in an infinite number of ways to create the diverse array of substances we encounter every day. Molecules of an element are formed when two or more atoms of that element join together. Two oxygen atoms are linked together to form oxygen gas ($O2$). These stable diatomic compounds can be found ubiquitously in the natural world.

What holds molecules together is chemical bonds.

Molecules are the fundamental building blocks of all matter, whereas atoms almost never exist alone. Chemical bonds, the "glue" that binds atoms in compounds, allow for these interactions to take place.

Covalent bonds and ionic bonds are the most frequent types of chemical bonds. Electron sharing is a key component of covalent bonding. Covalent bonds are the most stable type of bond between atoms because they involve the sharing of electron pairs. A water molecule (H_2O) is an example of a molecule having its own features resulting from the sharing of electrons between atoms of hydrogen and oxygen.

On the other hand, ionic bonds need the sharing of electrons between atoms. Ions, which can be charged positively or negatively, are created when electrons are transferred from one atom to another. Ionic compounds are formed when ions with opposite charges bind together. NaCl, or table salt, is a prototypical ionic compound because it consists of sodium ($Na+$) and chlorine ($Cl-$) ions that are bound together by electrostatic forces.

The huge variety of molecules is the result of the existence of covalent and ionic bonding. They determine how substances are put together and how they behave. To comprehend molecules and their roles in chemical reactions and the natural world, one must have a firm grasp of the nature of these relationships.

From the Most Basic to the Most Complicated Molecules

The molecular world is characterised by incredible variety. Different molecules, with varying degrees of complexity, serve different purposes.

Hydrogen (H_2), nitrogen (N_2), and oxygen (O_2) are all examples of diatomic molecules, which are made up of only two atoms of the same element. These chemicals play crucial roles in activities as diverse as respiration and combustion.

Some examples of inorganic molecules are hydrogen (H_2), oxygen (O_2), carbon dioxide (CO_2), and ammonia (NH_3). Particularly important for maintaining the integrity of biological systems, water is a precondition for all forms of life on Earth.
Carbon-containing organic molecules are the backbone of biology because of their ability to form covalent bonds with other elements including hydrogen, oxygen, nitrogen, sulphur, and phosphorus. Carbohydrates, lipids, proteins, and nucleic acids are all types of organic molecules. The metabolic processes that support life on Earth depend on these molecules, which play essential roles in cellular structure, energy storage, information transfer, and more.

Macromolecules are a further illustration of the complexity of molecules. These are very complex macromolecules made up of smaller molecules called monomers that repeat. Amino acid monomers are the building blocks of proteins, while peptide bonds hold the molecule together. It is the amino acid sequence and the protein's three-dimensional structure that define whether it is an enzyme that catalyses chemical reactions or a structural component of a cell.

Finally, a Universe Made of Molecules.

The fundamental principles of chemistry have been uncovered via the study of atoms and the elements that make up molecules. Elements are identified by the subatomic particles that make up their atoms. The periodic table of elements serves as chemistry's "alphabet." The world's incredible variety of substances is the result of chemical bonds, either covalent or ionic, which act as the forces that build molecules and compounds.

Complex molecules recount the stories of matter and life, and they do so as the narrators of chemistry. Molecules, from the tiniest diatomic particles to the largest biological macromolecules, are the building blocks of all known matter. Exploring them is a never-ending trip into the very heart of chemistry because they are physical manifestations of the marvellous diversity and unity that determine the composition of matter.

2.2- How atoms bond to form molecules.

Molecule formation and the atomic bonding process.

The building blocks of all matter, atoms, are like jigsaw puzzle pieces in space just waiting to be pieced together to form molecules. Chemistry is beautiful because it explains how atoms combine and link to form the myriad compounds that make up our universe. In this investigation, we delve into the intriguing realm of chemical bonding and learn how atoms are held together and molecules are formed.

What Drives Chemical Bonding and Why

Chemical bonds form when atoms try to maintain their stability by exchanging or giving up electrons. It is in the nature of atoms to strive for a stable electron configuration, taking cues from the very stable noble gases, which have entire outer electron shells.

The formation of chemical bonds between atoms is the means through which this stability is achieved. These connections, or bonds, between atoms allow them to join in a variety of configurations to form molecules. Covalent bonds and ionic bonds are the most frequent types of chemical bonds.

The Covalent Bond: An Example of Electron Sharing

It is through the sharing of electrons that covalent bonds are established between atoms. When two atoms have incomplete outer electron shells and can profit from sharing electrons, they do so.

Think about the prototypical molecule of water (H_2O). In this scenario, a molecule is formed from the combination of two hydrogen (H) atoms and one oxygen (O) atom. Helium, the noble gas with two electrons in its outermost shell, has a different electron configuration than hydrogen, which only has one. With its outermost shell already filled with six electrons, oxygen needs just two more to reach the stable configuration of neon.

Each hydrogen atom in water shares an electron with an oxygen atom, creating a covalent link. By exchanging electrons, all the atoms in the molecule will have their outermost shells completely filled. To create the stable H2O molecule, the oxygen atom gives up one of its electrons to each hydrogen atom.

The quantity of shared electrons determines the type of covalent bond formed. When two atoms form a covalent bond, like H2O does, they share two electrons. In molecules like oxygen gas (O2), four electrons are shared in a double covalent bond. Three atoms of nitrogen gas (N2) share six electrons in a triple covalent connection.

Covalent bonds are the most common type of chemical bond and their strength is determined by the amount of shared electrons and the separation of the nuclei of the atoms involved. Stable molecules are characterised by strong covalent bonds, the breaking of which requires a great deal of energy.

Electron Transfer in Ionic Bonds

However, ionic bonds are formed when electrons are shared between atoms. Due to electrostatic repulsion, ions with opposite charges are created during this exchange. Atoms with very varied electronegativities (the strength of their attraction to electrons) form ionic bonds with one another.

The production of sodium chloride (NaCl), or table salt, is a prototypical example of ionic bonding. Both sodium (Na) and chlorine (Cl) have one electron in their outermost shell, but chlorine has seven. When exposed to chlorine, sodium rapidly gives up its lone electron, turning both atoms into ions. When ionised, sodium takes on a positive charge (Na+) while chlorine takes on a negative charge (Cl-).

As a result of their electrostatic repulsion towards one another, these ions form an ionic bond that keeps the complex together. The solid salt we use in our kitchens is formed when sodium and chlorine ions are organised in a repeating pattern in a crystal lattice.

Metal-organic complexes are common examples of ionic bonding. Ions (cations) are formed when a metal loses an electron, while ions (anions) are formed when a nonmetal gains an electron. Ionic compounds are formed when ions with opposite charges are drawn together by electrostatic forces.

Electron Sharing in Polar Covalent Bonds

The sharing of electrons in covalent bonds is not always equimolar. Polar covalent bonds develop when electrons are shared inequitably between two atoms. Electronegativity-dependent bonds form when one atom exerts a stronger pull on the shared electrons than the other atom.

Hydrogen fluoride (HF) is a chemical that exhibits a polar covalent bond. Because it is more electronegative than hydrogen, fluorine (F) has a stronger attraction for the shared electrons. This causes the electrons to spend more time in its vicinity, giving the fluorine atom a partial negative charge (-) and the hydrogen atom a partial positive charge (+). This charge separation results in a polar molecule with a more negatively charged end and a more positively charged end.

The chemistry of water relies heavily on polar covalent bonding. H_2O is a polar molecule with a partial negative charge on the oxygen and partial positive charges on the hydrogens because the oxygen atom is more electronegative than the hydrogen atoms. Water's ability to form hydrogen bonds and its high surface tension are two examples of the special features made possible by its polarity.

Hydrogen bonds: polar molecule attraction

Hydrogen bonds are a distinct form of chemical connection that develop between molecules with opposite charges. In contrast to covalent or ionic interactions, which are real chemical bonds, these intermolecular forces are quite powerful.

Hydrogen atoms that are already covalently bound to highly electronegative atoms (usually nitrogen, oxygen, or fluorine) attract and join with other electronegative atoms in neighbouring molecules to establish these bonds. Hydrogen atoms in one molecule tend to gravitate towards the more electronegative atoms in other molecules because of their shared attraction to hydrogen.

Hydrogen bonding can be seen in action in water. The partial positive charges on the hydrogen atoms and the partial negative charges on the oxygen atom are the result of the covalent bonding between the hydrogen and oxygen atoms in a water molecule. Water's unusual qualities, including its high boiling and melting points, its ability to dissolve a wide range of compounds, and its surface tension, are the result of strong interactions between neighbouring water molecules, which are a result of these partial charges.

Hydrogen bonds are critical in living systems because they determine how molecules like DNA and proteins are structured and what functions they perform. Their contributions to the molecular behaviour and compound creation in biological processes are equally substantial.

Intermolecular Attractions Due to Van der Waals Forces.

There are more types of intermolecular interactions besides covalent bonds, ionic bonds, and hydrogen bonds; these are called van der Waals forces. These pulls and pushes originate from transient positive and negative charges brought about by random shifts in electron distribution inside molecules. Dipole-dipole interactions and the London dispersion force are both examples of Van der Waals forces.

When two polar molecules contact, they engage in dipole-dipole interactions, in which the positive end of one molecule is drawn to the negative end of the other. Substances like hydrogen chloride (HCl) are stable because of these interactions, which occur between the partially positive hydrogen charge and the partially negative chlorine charge.

In contrast, the interactions between nonpolar molecules, known as London dispersion forces, are the smallest intermolecular forces. Temporary fluctuations are responsible for these forces.

, can cause transient positive and negative charges by altering the distribution of electrons. The electron cloud can become temporarily unevenly distributed, creating positive and negative patches, even in nonpolar molecules like methane (CH_4). These transient charges can lead to attraction between two nonpolar molecules, which in turn can lead to London dispersion forces.

Atomic bonding: a climactic conclusion.

In chemistry, the bonding forces that hold atoms together perform a mesmerising dance. The chemistry of bonding is a complex web of interactions that ranges from the sharing of electrons in covalent bonds to the transfer of electrons in ionic bonds, from the polarity of molecules to the delicate attractions of hydrogen bonds and van der Waals forces.

From the simplest diatomic gases to the most complex macromolecules of life, these bonds and forces serve as the imaginative threads that hold molecules together. They characterise what substances are like, how they behave, and what they do. Aside from being a cornerstone of chemistry, the process by which atoms connect to create molecules is crucial to understanding the workings of the natural world and tapping into its potential for scientific advancement and technological advancement.

Chapter 3.
Water, the Life-Giver:

3.1- A deep dive into the unique properties of water and its importance in biology.

An In-Depth Exploration of Water's Biological Importance and Its Special Characteristics

It's no exaggeration to say that life as we know it couldn't exist without water, the "universal solvent," a fascinating substance with a complicated and unique set of qualities. It is a molecule that drives the basic mechanisms that allow living things to survive and grow. This investigation will delve deeply into water's amazing qualities and its crucial function in the living world.

Water's Structure

Two hydrogen atoms (H) are covalently connected to one oxygen atom (O) at the centre of a water molecule, creating a V-shape. There is an unequal distribution of electrical charge within the molecule because the oxygen atom attracts electrons more strongly than the hydrogen atoms.

Since the oxygen atom ends up with a negative charge (-) and the hydrogen atoms end up with positive charges (+), the resulting covalent bond is polar. Many of water's special qualities can be traced back to this polarity.

Polarity and hydrogen bonding are both quite strong.

Hydrogen bonding is a unique characteristic of water that arises from its polarity. In water, the positively charged hydrogen atoms of one molecule attract the negatively charged oxygen atom of another through weak electrostatic attractions known as hydrogen bonds.

Each water molecule is continually making and breaking hydrogen bonds with its neighbours. The high heat capacity, high surface tension, and peculiar density behaviour of water are all gifts from this network.

Able to hold a lot of heat

The large heat capacity of water is essential for the regulation of temperatures in aquatic ecosystems and living things. Water's vast hydrogen bonding network allows it to absorb and store large amounts of heat energy without significantly raising its temperature. It helps to smooth out temperature swings in water bodies like the ocean, lake, and even the human body.

Because of this quality, living things can adapt to different temperatures and continue to function normally. The consistent temperatures in bodies of water, for instance, provide environments in which aquatic creatures can thrive throughout the year.

Surface tension is very high.

Hydrogen bonding in water also contributes to its high surface tension. Surface tension is the force required to break the liquid's surface. Because of their mutual attraction, water molecules at the surface form a tough, impenetrable "skin" that protects the underneath layer. This explains why bugs and leaves of a certain size can sometimes seem to "float" on the surface of a body of water.

Some aquatic species, such as insects that can walk on water or plants that employ capillary action to pull water from the ground to their leaves, rely on a high surface tension for diverse activities.

Strange Density Phenomena

In general, the density of a substance increases when it cools and decreases when it warms. Water, however, acts in a peculiar manner. Its maximum density is reached at a temperature just beyond its freezing point, around 4 degrees Celsius (39 degrees Fahrenheit). When water is cooled to a temperature below 4 °C, it expands and loses density. Because of this, ice floats on water.

This odd quality is essential for survival in frigid environments. The insulating layer of ice that forms on the surface of a body of water after freezing keeps the water underneath from also becoming solid. This makes it possible for aquatic species living in the liquid water beneath the ice to survive the harsh winters.

Properties that make a solvent

A large variety of chemicals can be dissolved in water because of its remarkable solvent capabilities, making water a universal solvent. Water's polarity means it may bind to and enclose ions and polar molecules, effectively isolating and distributing them over a solution.

In biology, the ability to dissolve other molecules is essential for the transfer of essential nutrients, ions, and other molecules throughout the body. Water is essential for the dissolution and movement of important components like amino acids, carbohydrates, and ions within cells, and it also acts as the medium in which biochemical events take place.

Hydrophilic molecules and the process of hydration.

Hydration refers to the process through which water forms bonds with other molecules, often those of solutes that are both polar and ionic. Particles in solution are protected from their charges by the water molecules that surround them. Proteins and nucleic acids, two examples of biological macromolecules, require water to maintain their structure and activity.

The polar or charged character of hydrophilic molecules makes them easily soluble in water. Sugars, salts, and amino acids are all examples of hydrophilic compounds found in living organisms because of their ability to form hydrogen bonds with water. This quality permits these molecules to move about inside living things and to take part in biological processes.

The Importance of Hydrophobic Interactions in Hydrophobic Molecules

Hydrophobic molecules, in contrast to hydrophilic ones, tend to cluster in aqueous solutions because they are repulsed by water. When these nonpolar molecules are pulled together by the surrounding water, they create an environment in which they can minimise contact with water molecules, an effect known as hydrophobic interactions.

The folding of proteins and the assembly of lipid bilayers in cell membranes are two examples of where hydrophobic interactions are crucial. To aid in protein folding and membrane formation, water is kept away from the hydrophobic core of proteins and the hydrophobic tails of phospholipids in cell membranes.

The Importance of Water in Living Systems

There are numerous biochemical events and physiological functions within living creatures in which water plays a crucial and active role.

Reactions in Biology

Water is frequently involved as either a reactant or a product in biological reactions. In the chemical reaction known as hydrolysis, for instance, water is utilised to hydrolyze (or split) complex molecules into simpler ones. In the digestive process, water is used by enzymes to break down macromolecules like carbohydrates, proteins, and lipids into their component monomers.

In contrast, condensation reactions, also known as dehydration synthesis, involve water removal to join monomers into bigger molecules. Synthesis of nucleic acids and proteins, two of the most complicated biological components, relies heavily on this procedure.

Controlling the Temperature

Water is an efficient coolant and temperature regulator in living systems due to its high heat capacity and heat of vaporisation. Evaporation of water from the surfaces of organisms, especially those living in aquatic habitats, is one way they can dissipate excess heat. This procedure aids in regulating core body temperature and wards from overheating.

Nutrient and waste transportation

Transport of nutrients, ions, and waste materials within and between cells relies on water. For instance, blood is mostly water and transports oxygen, nutrients, and waste materials from and to cells all over the body. Plants are able to carry water and nutrients thanks to the water movement in their vascular tissues, such as xylem and phloem.

Lubrication

The lubricating effects of water are crucial to many bodily processes. Synovial fluid, which is mostly water, functions as a lubricant in joints to prevent bones from rubbing against one another and to facilitate easy, painless motion. Similar to how water in the mucus of the respiratory and digestive systems acts as a lubricant to protect and allow for the movement of sensitive tissues, water is also present in the mucus of the reproductive system.

Water is the "elixir of life," the conclusion reads.

Water is a very unusual molecule because of its diverse properties and pivotal function in all forms of life. It's the stuff that keeps everything alive, the elixir of the gods.

, which is what powers all living things. Water's role as a temperature regulator and as a solvent in biochemical reactions make it an essential substance in the natural world.

There is no place on Earth where water's impact is more pervasive than at the depths of the oceans, the interiors of living cells, and the complex web of life. It is the unsung hero of the biological symphony, contributing to the well-being and development of all forms of life. Appreciating the wonders of the natural world around us requires not only a technical understanding of water's fundamental properties and importance.

3.2- How water molecules shape the biological world.

How the molecules of water influence the living world.

The universal solvent, water, is the true unsung hero of life. Its extraordinary qualities and peculiar habits have shaped the terrains of Earth's biosphere and continue to determine the fate of innumerable living things. Water molecules are the medium through which the dance of biology is performed, from the deepest parts of the ocean to the tiniest parts of our bodies. In this investigation, we dig into the amazing role that water molecules play in moulding the biosphere.

Water's Multipurpose Nature.

Water is a straightforward molecule with just two H atoms and one O atom held together by covalent connections. However, water's complicated behaviour belies its simple molecular composition. Water is both a necessary component of life and a unique liquid with many interesting properties.

Water's Polar Origins

One of water's defining characteristics is its polarity. Water molecules are polar because oxygen and hydrogen share electrons unequally, leading to a partial negative charge (-) on the oxygen atom and a partial positive charge (+) on the hydrogen atom. Because of its polarity, the molecule's electrical charge is unevenly distributed.

Water's polarity is responsible for many of its remarkable qualities, such as its solubility of a wide variety of compounds and its ability to create hydrogen bonds. The importance of these hydrogen bonding in the way water affects living things cannot be overstated.

Hydrogen bonds are the building blocks of life.

In water, the positively charged hydrogen atoms of one molecule attract the negatively charged oxygen atom of another through weak electrostatic attractions known as hydrogen bonds. Due to the constant mobility of water molecules, these connections are extremely dynamic, constantly forming and breaking. Many of water's extraordinary features stem from the dynamic network of hydrogen bonding.

Able to hold a lot of heat

Water has a high heat capacity because it can absorb and store heat energy without significantly raising its temperature. The vast hydrogen bonding network in liquid water is responsible for this characteristic. When you heat water, some of that heat goes towards breaking hydrogen bonds before the water's temperature rises.

The ability to maintain a constant internal temperature and adapt to varying ambient temperatures necessitates a high heat capacity, making it essential for all forms of life. Oceans and lakes in particular serve as thermal buffers, moderating the local climate and maintaining a consistent environment for aquatic life.

Extremely High Vaporisation Temperature

The amount of energy needed to turn a certain amount of liquid water into vapour at room temperature is called the heat of vaporisation, and water has a high heat of vaporisation. Because of this quality, water functions well as a biocoolant in living systems. Plants and animals alike use the heat from their surroundings to evaporate water through perspiration and transpiration, respectively.

This chilling action is critical for keeping internal body temperatures within a narrow range. In hot climates, life forms would quickly overheat without water's high heat of vaporisation.

Abnormality in Density

Water has a peculiar density behaviour compared to other substances. Its maximum density is reached at a temperature just beyond its freezing point, around 4 degrees Celsius (39 degrees Fahrenheit). When water is cooled to a temperature below 4 °C, it expands and loses density. Because of its special quality, ice can be seen floating on the surface of a body of water.

Water's density discrepancy has far-reaching effects on marine life. The ice that forms on the surface of frozen bodies of water insulates the water below from further freezing. The liquid water beneath the ice provides a safe haven for aquatic species throughout the cold winter months.

The Key to Eternal Happiness

The solvent qualities of water are crucial to all forms of life. It can dissolve so many different materials that it has earned the nickname "universal solvent." Water's polarity gives it the ability to surround and interact with ions and polar molecules, thereby isolating them and allowing them to disperse uniformly throughout the solution.

Many biological activities rely on water's solubilizing properties. Water is the medium for metabolic reactions within cells. Ions, amino acids, carbohydrates, and nucleic acids all rely on it to dissolve, be transported, and undergo chemical reactions.

Stability of Biological Macromolecules in the Presence of Water.

Hydration refers to the process by which water interacts with solute molecules, especially polar and ionic substances. Solute particles are protected from their charges and kept in solution by a layer of water molecules. The integrity and functionality of biological macromolecules depend on this mechanism.

Hydration is essential for the structure and function of many molecules, including proteins. Proteins are able to fulfil their biological tasks because of the unique ways in which their hydrophilic (water-loving) and hydrophobic (water-hating) regions interact with water and other molecules.

Molecular Hydrophiles and Hydrophobes

The polar or charged character of hydrophilic molecules makes them easily soluble in water. Sugars, salts, and amino acids are all examples of hydrophilic compounds found in living organisms because of their ability to form hydrogen bonds with water. Because of this quality, these molecules can move about inside living things and take part in metabolic reactions.

In contrast, water repulsion causes hydrophobic molecules to cluster in liquids. When nonpolar molecules are pushed closer together by the surrounding water, a hydrophobic interaction occurs, allowing the molecules to have as little contact with water as possible.

Structure and Function of Membranes

The creation and integrity of cell membranes depend critically on water's interactions with hydrophobic substances. Phospholipids are lipid molecules found in cell membranes; they are characterised by a hydrophilic phosphate head and a hydrophobic lipid tail.

Phospholipids form a bilayer structure in water, with the water-loving hydrophilic heads on the outside and the water-hating hydrophobic tails on the inside. This lipid bilayer is the structural backbone of cell membranes, acting as a selective barrier to regulate chemical transport across cell boundaries.

The Importance of Water in Living Systems

Many biochemical reactions and physiological functions within living beings rely on water, making it clear that water is not a bystander to biological processes.

Controlling the Temperature

Water is an efficient coolant and temperature regulator in living systems due to its high heat capacity and heat of vaporisation. Water's heat-absorbing qualities help maintain ideal body temperatures and prevent overheating; organisms can release surplus heat through processes like sweating or transpiration in plants.

Transportation of both nutrients and waste

Transport of nutrients, ions, and waste materials within and between cells relies on water. The water-based blood transports oxygen, nutrients, and waste materials to and from the body's cells. Water moves through a plant's vascular tissues, allowing nutrients and water to be transported from the roots to the leaves.

Lubrication

The lubricating effects of water are crucial to many bodily processes. Synovial fluid is mostly water and functions as a lubricant in joints to prevent painful grinding of bones against one another. Similar to how water in the mucus of the respiratory and digestive systems acts as a lubricant to protect and allow for the movement of sensitive tissues, water is also present in the mucus of the reproductive system.

Final Thoughts: Water, the Master Builder of Life

In the living world, water molecules play the role of chief engineers. Due to factors such as polarity, hydrogen bonding, and high heat,

 capability and financial stability, have permanently altered the biota of our world. Water is the canvas onto which the exquisite patterns of biology are painted, the medium through which the symphony of life is performed.

Water's impact ranges from the microscopic to the planetary, from the rustle of a leaf to the thunder of a waterfall. It is an indispensable component of the biological universe since it shapes the conditions in which organisms exist, controls the processes of life, and so on.

To learn how water molecules shape the biological world is to have a deeper understanding of the interdependence of all living things and the complex dance of molecules that keeps Earth habitable.

Chapter 4.
The Energy Molecules, Carbohydrates

4.1- Understanding carbohydrates and their role as energy sources.

The Importance of Understanding Carbohydrates as a Source of Energy

Carbohydrates, or "carbs," are a type of macronutrient along with protein and fat that are necessary for human survival. These molecules are not only necessary for survival, but also provide the body with its primary source of energy. The structure, functions, and essential significance as energy sources will be uncovered as we delve into the world of carbs in this investigation.

Fundamentals of Carbohydrates

Carbohydrates are organic molecules with a common atomic makeup of two carbons, two hydrogens, and one oxygen. The word "carbohydrate" itself gives away their chemical make-up, consisting of the prefix "carbo" for carbon and the suffix "hydrate" for water. Carbohydrates get their name because their carbon atoms are frequently hydrated, or bonded to water molecules.

There is a wide spectrum of carbohydrates, from the simplest sugars to the most intricate polysaccharides. Single-sugar molecules, or monosaccharides, are the simplest type of carbohydrate. Glycose, fructose, and galactose are three often encountered instances. These monosaccharides are the basic material from which complex carbs are constructed.

Two monosaccharides are joined together to form a disaccharide. Examples of disaccharides are sucrose (made up of glucose and fructose) and lactose (made up of glucose and galactose).

Disaccharides, which are plentiful in food, are a quick and easy way to get energy.

Instead, polysaccharides are lengthy chains of simpler sugars called monosaccharides. Polysaccharides like starch and glycogen are prevalent in the human diet. When it comes to storing energy, plants use starch, while animals and humans rely primarily on glycogen.

Carbohydrate Functions

Beyond functioning as a source of fuel, carbohydrates have many other vital functions in the body. Their primary roles consist of:

1. Power Plant The majority of our daily calories come from carbohydrates. They are metabolised in the body to glucose, which can be utilised right away, or stored as glycogen in the liver and muscles for later use.

Supporting Structures: Carbohydrates play a crucial role as structural molecules in all forms of life. Cell walls in plants are made of cellulose, a complex carbohydrate that gives the cells strength and stiffness. Chitin, another carbohydrate, makes up a large portion of an arthropod's exoskeleton.

Thirdly, Storage: Carbohydrates are used for energy storage. When more energy is needed, the body can quickly convert the glycogen stored in the liver and muscles into glucose.

[4] [Transportation] Nucleic acids (RNA and DNA) are the molecules responsible for storing and transmitting genetic information.

The immune system is responsible for recognising and responding to foreign substances thanks to glycoproteins, which are carbohydrates attached to proteins. They function in cell signalling, adhesion, and interactions between immune cells.

Digestive Health #6: Dietary fibre is an indigestible carbohydrate present in plant diets that is needed for healthy bowel function. It encourages frequent defecation, keeps gut flora in good shape, and may lessen the likelihood of several digestive diseases.

Energy from carbohydrates

Carbohydrates' principal function is as a source of fuel for the body. Let's take a look at glucose metabolism to see how this works:

(1) Digestive system: The digestive tract is the starting point, when dietary carbohydrates are metabolised into simpler sugars. Carbohydrate digestion involves enzymes breaking up the chemical connections between sugar molecules in the mouth, stomach, and small intestine.

Absorption, number 2. Glucose and other simple sugars are absorbed into the bloodstream via the intestinal walls after being digested. They are then distributed to the numerous organs and tissues of the body.

3. Respiration in Cells Cells use glucose as a fuel source through a process called cellular respiration, where it is converted into adenosine triphosphate (ATP). Energy from glucose oxidation is stored as ATP during cellular respiration.

The equation for cellular respiration is simplified as follows:

Energy (ATP) is produced when glucose is combined with oxygen.

Energy is generated by storing it in ATP molecules and releasing it when cells need it. Muscle contraction, neuron signalling, protein synthesis, and other vital biological processes all rely on this energy.

Storage 5. Glycogen is a storage form of glucose that can be found in the liver and skeletal muscles. When glucose levels in the blood drop, the body converts glycogen to glucose and releases it into the bloodstream.

Carbohydrates in the Diet.

Carbohydrates can be found in a variety of food groups, from grains and cereals to fruits and vegetables to legumes to dairy to sugary treats. There are two main types of carbohydrates, simple sugars and complex sugars, that describe these food components.

Sugars and other simple carbohydrates: One or two sugar molecules make up a simple carbohydrate, sometimes known as "sugars." Sucrose (table sugar), which consists of one glucose molecule and one fructose molecule, is also a simple carbohydrate. Blood sugar levels can quickly rise after consuming simple carbohydrates because of how quickly they are processed.

The Role of Complex Carbohydrates: Long chains of sugar molecules make up complex carbohydrates, which are also known as "starches." Grains (such rice, wheat, and oats), legumes (like beans and lentils), and starchy vegetables (like potatoes and corn) are all examples of complex carbs. Complex carbs provide a steady stream of energy since they are metabolised more slowly than simple carbohydrates.

Nutritional Value of Carbohydrates

One's health can be greatly affected by the type of carbs they consume. Energy and nutrients can be sustained throughout the day by eating a diet high in whole, unprocessed foods such grains, fruits, vegetables, and legumes. In addition to boosting digestive health and lowering the risk of chronic diseases like heart disease and type

2 diabetes, the dietary fibre that typically accompanies these carbohydrates offers its own set of health benefits.

Blood sugar levels can fluctuate rapidly in response to a diet that is high in added sugars, refined carbohydrates, and processed meals.

, which could lead to obesity, insulin resistance, and metabolic disease.

Special Dietary Carbohydrates

A person's carbohydrate consumption should be tailored to their own dietary preferences and requirements. The goal of the low-carb and ketogenic diets, for example, is to minimise carbohydrate intake and increase fat burning for energy. These diets have risen in favour as a means of both losing weight and controlling the symptoms of diseases like epilepsy.

Carbohydrate-heavy diets, like the high-carbohydrate diets popular among endurance athletes, give an instantaneous source of energy to fuel exercise.

Eating a Healthy Amount of Carbohydrates

Most people do well on a diet that incorporates a wide range of carbohydrates from a variety of complete food sources. A diet like that supplies the fuel you need to get through the day and is good for your health in general. The proportion of carbs, proteins, and fats in a person's diet can be altered to achieve various health and fitness objectives, but carbohydrates will always play a pivotal function in the body as a primary energy source.

Carbohydrates: The Body's Main Source of Energy

Carbohydrates are not optional; they provide the body with a necessary source of fuel. Carbohydrates, found in everything from simple sugars to complex starches, are essential for providing energy for our daily activities and keeping our cells alive and well.

Individuals may better support their health through their diet when they have a firm grasp on the many forms of carbohydrates, how they are metabolised, and the effects they have on the body. Carbohydrates are essential to human nutrition because they shape our bodies and provide the energy we need to do anything from play sports to live healthy lives.

4.2- Glycolysis and the chemistry of glucose.

The Chemistry of Glucose and the Glycolytic Process

Glycolysis is the star biochemical route in the complex show that is cellular metabolism. Glycolysis can be broken down into a sequence of chemical reactions that release energy by decomposing glucose, a simple sugar, into smaller molecules. The chemistry of glucose and its transformation via glycolysis are key to comprehending this essential route.

Glucose Chemistry

Glucose, or blood sugar, is a sugar composed of six carbon atoms and has the formula $C_6H_{12}O_6$. Its ability to provide an essential source of energy in living creatures may be traced back in large part to its hexagonal structure, which is known as a hexose. To make the chemical change of glucose easier to describe, we've assigned numbers from 1 to 6 to each of its carbon atoms.

The Carbon Structure of Glucose:

The First Carbon (C1): A carbonyl group (C=O) includes this carbon atom, making it a member of the aldehyde functional group. Glycolysis depends on glucose's reactivity, which is provided by its aldehyde group.

C2 is the second carbon atom in the sugar molecule's backbone.

Carbon 3 (C3): The third carbon is a component of the glucose backbone where it generates a hydroxyl group (-OH).

Glucose's hexagonal structure is due in part to the presence of carbon-4 (C4), another carbon found in the backbone.

Fifth carbon (C5): C5 is present in the backbone of glucose.

Carbon 6 (C6), number 6. The sixth carbon in glucose's molecule is represented by the number C6.

Formation of Rings and Isomers:

Glucose in water takes the form of a ring rather than a linear chain. The alpha and beta anomers of this ring structure are possible. The hydroxyl group on Carbon 1 is oriented differently in the two variants. The hydroxyl group is oriented downward in the alpha anomer and upward in the beta anomer. Biological reactions involving glucose are greatly affected by this isomerism.

Energy Production through Glucose Catabolism

The term "glycolysis," which originates from the Greek words "glykys" (sweet) and "lysis" (splitting), describes the process by which glucose is broken down into its component parts. The first step in both aerobic (with oxygen present) and anaerobic (without oxygen present) cellular respiration, this route takes place in the cytoplasm of cells.

There are eleven stages of glycolysis, all of which require enzymes to complete. Let's take a closer look at the major processes that illustrate the chemistry of glucose throughout glycolysis:

Glucose phosphorylation is the first step.

The first step involves adding a phosphate group to glucose. Hexokinase is an enzyme that catalyses this process, which takes one molecule of ATP. The conversion of glucose to glucose-6-phosphate keeps glucose within the cell and blocks its movement across the membrane.

Second, fructose-6-phosphate is formed through isomerization.

Glucose-6-phosphate undergoes an isomerization event to become fructose-6-phosphate. This transformation is catalysed by the enzyme glucose-6-phosphate isomerase and results in a different atomic configuration but the same chemical formula.

Thirdly, fructose-6-phosphate is phosphorylated.

In the presence of ATP, fructose-6-phosphate is converted into fructose-1,6-bisphosphate. Phosphofructokinase is the enzyme responsible for catalysing this process, which also requires the presence of ATP.

4. Fructose-1,6-Bisphosphate Cleavage

By hydrolysis, fructose-1,6-bisphosphate is split into dihydroxyacetone phosphate (DHAP) and glyceraldehyde-3-phosphate (G3P), both of which contain three carbons. These isomers can freely transform into one another. This breakdown is catalysed by the enzyme aldolase.

Production of NADH and Energy, Steps 5-9

Through a sequence of processes, G3P is transformed into 1,3-bisphosphoglycerate (1,3-BPG), while also producing ATP and NADH. For every molecule of glucose that undergoes glycolysis, two ATP molecules are produced.

Tenth Step: ATP Synthesis

The final phase involves converting 1,3-BPG to pyruvate, where two molecules of ATP are generated for every molecule of glucose. Pyruvate kinase is the enzyme responsible for catalysing this process.

Pyruvate's Uncertain Future

Glycolysis culminates in the formation of pyruvate, a three-carbon complex consisting of two glucose molecules. What happens to pyruvate depends on whether or not oxygen is present:

To create extra energy through oxidative phosphorylation, pyruvate enters the citric acid cycle (Krebs cycle) when oxygen is present.

Depending on the organism, pyruvate can either be transformed into lactate or fermented into alcohol in the absence of oxygen. Because of this, NAD+ molecules can be replenished, allowing glycolysis to proceed even when oxygen is unavailable.

Glycolysis Regulation

During times of high energy demand, glycolysis is strictly regulated to ensure that glucose is transformed into usable cellular fuel. Allosteric regulation occurs at multiple enzymes involved in glycolysis, where particular molecules bind to regulatory regions on the enzyme and alter its activity. When energy levels are high, for instance, ATP and citrate suppress the action of phosphofructokinase, a critical enzyme in glycolysis.

Hormonal cues like insulin and glucagon react to shifts in blood glucose levels to modulate glycolysis. Glucagon raises blood glucose levels and inhibits glycolysis, while insulin enhances glucose uptake by cells and stimulates glycolysis.

Glycolysis: The First Step in Energy Production

Glycolysis is the biochemical pathway via which cells generate energy. The complex chemistry involved in this process converts glucose, a simple sugar, into ATP and NADH, energy-rich molecules essential to cellular function. Grasping the fundamental processes that support life and drive cellular metabolism requires an understanding of the chemistry of glucose and its path through glycolysis. The accuracy and intricacy of biochemical processes are what give life its vibrancy, and this pathway's chain of enzyme reactions and regulatory systems is a prime example of this.

Chapter 5.
Lipids, or Life Fats,

5.1- Examining the structure and functions of lipids in biological systems.

Lipids: Analysing Their Structure and Biological Roles

Organic molecules known as lipids have several important functions in living organisms. Lipids play numerous crucial roles in all forms of life, from facilitating communication between cells to storing energy. The complex structure and biological roles of lipids are the focus of this investigation.

The Wide-Ranging World of Lipids

Insoluble in water but soluble in nonpolar solvents like ether or chloroform, lipids make up a large class of hydrophobic chemical compounds. They cover a vast range of substances with different structures and roles, such as fats, oils, phospholipids, and steroids.

Commonalities exist among lipids notwithstanding their chemical variety:

Hypophobic Ending: Hydrocarbon chains or rings are the primary building blocks of lipids; they are nonpolar and hydrophobic. Because of this property, lipids tend to cluster in nonpolar surroundings and repel water.

Hydrophilic Noggin': A phosphate group or polar area are two examples of hydrophilic (water-attracting) components found in many lipids. Lipids can interact with water at interfaces thanks to their hydrophilic head, which allows them to form complex structures like cell membranes.

Power Reserves: Lipids' principal role is in energy storage. Lipids are an effective long-term energy reserve because their metabolism produces more energy per gramme than either carbs or proteins.

- Role in the Structure: Cell membranes, which encapsulate the cell and its various subcompartments, are largely composed of lipids and rely on them for their structure and function. Lipids' hydrophobic tails act as a gatekeeper by blocking the way for polar molecules, regulating what goes in and what comes out of cells.

The Most Typical Lipids

Here we will examine some of the most common lipid types in order to better comprehend the structure and roles of lipids:

1. Fatty Acids and Triglycerides Triglycerides, the most common type of dietary lipid, are used by animals and plants to store energy. A glycerol molecule joins three fatty acid chains to form its structure. The structural features of fats and oils are modified by the fatty acids they contain, which range in length and saturation. The lack of double bonds between carbon atoms in saturated fats makes them solid at room temperature, while the presence of double bonds in unsaturated fats makes them liquid (oils) at that temperature.

Second, phospholipids: Cell membranes can't function without phospholipids. Similar to triglycerides in structure, but with only two fatty acid chains and a phosphate group. Hydrophilic at the phosphate head, hydrophobic at the fatty acid tails. Phospholipids form the lipid bilayer of cell membranes in an aqueous environment, with the hydrophilic heads facing outside and the hydrophobic tails facing within.

(3) Anabolic Steroids: The lipid class known as steroids is characterised by its characteristic four-ring shape. Cholesterol is an essential steroid found in the membranes of all mammalian cells. The

fluidity and stability of membranes are helped by this. Steroids also play a role in the production of hormones like oestrogen, testosterone, and cortisol.

4. Waxes: Waxes are a type of water-repellent lipid present in many organisms. A long-chain fatty acid is joined to a long-chain alcohol to form these compounds. The waxy covering on plant leaves and the wax found in animal fur and feathers both serve to keep moisture out and provide insulation for the organism.

Lipids' roles in living organisms

The structural diversity and versatility of lipids are reflected in the breadth of their biological roles. Some of lipids' most important functions include the following:

Storage of Energy: Fats and oils (triglycerides) are a crucial source of stored energy for living things. These molecules are necessary for continued existence since they can be broken down in the metabolic process to provide energy when required.

2. Components of the Structure: Cell membranes can't function without lipids. The phospholipid-based lipid bilayer forms a selective barrier that regulates what can enter and exit cells. This membrane architecture is essential for steady cellular function and integrity.

(3) Soundproofing and Safety: Insulation and protection are provided by adipose tissue, which is made up of fat cells, in animals. Subcutaneous fat provides insulation, which is important for maintaining a healthy body temperature and warding off injury.

Signalling and regulation, number four: Cholesterol is the precursor to steroid hormones, which have crucial roles in signalling and regulation. Physiological functions such as metabolism, growth, and immunological response are all regulated by hormones like cortisol, oestrogen, and testosterone.

Fifthly, defence and watertightness: Waxy substances on plant leaves and stems protect them from infections and herbivores while also reducing water loss. Sebum, a waxy secretion found in animals,

 generated by sebaceous glands aids in keeping the skin hydrated and protected from harmful bacteria and fungi.

Sixth, transport: Hydrophobic lipids (such cholesterol and triglycerides) are transported through the circulatory system by lipoproteins, which are complexes of lipids and proteins. Cholesterol transport and metabolism rely heavily on high-density lipoproteins (HDLs) and low-density lipoproteins (LDLs).

Vitamin Absorption, Number Seven Vitamins A, D, E, and K are called fat-soluble vitamins because they are absorbed in the digestive tract with the help of certain fats. These vitamins can't be absorbed or transported within the body without the help of dietary lipids.

Metabolic Fuel, Number Eight In times of fasting or low carbohydrate intake, the body can turn to lipids for metabolic fuel instead of its usual source, glucose. When glucose is in short supply, fatty acids can be oxidised to produce ATP.

9. Variation in Membrane Structure The fluidity and functioning of membranes are affected by the variety of lipid types present, such as saturated and unsaturated fatty acids. Cells are able to keep their membranes functioning at their best regardless of whether or not the surrounding temperature or environment is ideal.

Lipids, the Multipurpose Molecules

Because of their structural and functional diversity, lipids are often called "the versatile workhorses of biology." Lipids serve crucial roles in the intricate web of biological activities, including energy storage, membrane formation, and hormone signalling.

The importance of lipids in protecting cells and organisms, adjusting to different circumstances, and allowing for the diversity of life forms to flourish can be better understood if we take the time to learn about their structure and function. Lipids play crucial roles in all life on Earth, whether as energy stores, protective barriers, or signalling molecules.

5.2- Lipid membranes and their significance.

Importance of Lipid Membranes in the Body

The presence of lipid membranes is essential for all forms of life. Lipid bilayers are the dynamic, double-layered structures that surround cells and organelles and are responsible for maintaining cellular integrity and directing chemical traffic. Lipid membranes are critically important in biology since they not only serve as a physical barrier but also in activities such as signalling, adaptation, and cellular communication. In this investigation, we will delve into the fascinating world of lipid membranes and learn about their many roles in biology.

The lipid bilayer is the structural basis for cellular membranes.

Lipid bilayers are the primary building block of lipid membranes. Each lipid molecule has two tails, one of which is hydrophilic (attracts water) and the other hydrophobic (repels water), forming two layers. The hydrophilic heads are directed outward to engage with the surrounding water, while the hydrophobic tails are tucked in and out of the way.

Bilayer lipid membrane composition:

First, phospholipids: Lipid bilayers are made primarily of phospholipids. They are made up of two hydrophobic fatty acid tails and a hydrophilic head made up of a phosphate group and a glycerol molecule. In order to form bilayers, phospholipids must have both hydrophilic and hydrophobic areas.

Two words: cholesterol. The phospholipid bilayer contains pockets of cholesterol. They help keep the fatty acid tails from packing too tightly together, which is important for membrane fluidity and stability.

[Proteins] 3. The lipid bilayer is related with or contains membrane proteins. They can either be embedded in the membrane (integral) or attached to its surface (peripheral). Transport, signalling, and structural support are just a few of the many roles that membrane proteins play.

[4] [[Carbohydrates]] On the membrane's outer surface, carbohydrates can link to lipids or proteins. These carbohydrate chains play an important role in immunological responses, as they facilitate cell identification and adhesion.

Lipid membranes' biological significance

The lipid membranes that surround cells and cellular organelles play a crucial role in living organisms. Lipid membranes are dynamic, functional structures that underpin various essential biological processes and have far-reaching implications beyond those of simple physical barriers.

1) The Maintenance of Cellular Purity and Molecular Separation:

Cellular membranes are made up of lipids and act as selective barriers between the inside of the cell and the outside. This partition helps keep cellular components safe and functional throughout intracellular operations.

Lipid membranes separate the many organelles within a cell, including the nucleus, endoplasmic reticulum, mitochondria, and lysosomes. This compartmentalization of incompatible metabolic activities into their own membrane-bounded organelles is what makes cellular specialisation possible.

Transport inside Cells

Controlling how chemicals enter and exit cells is facilitated by lipid membranes. Some compounds can pass through them while others can't because of their selective permeability. In order to keep the internal environment stable for cellular functions, selective permeability is essential.

Diffusion and assisted diffusion are examples of passive transport methods that rely on the selective permeability of the lipid bilayer. Membrane proteins, which are themselves embedded inside the lipid bilayer, are responsible for active transport mechanisms that require the consumption of energy (often in the form of ATP).

Third, Cellular Recognition and Signalling:

When it comes to cell signalling and recognition, lipid membranes are crucial. The regulation of cellular responses is initiated by the interaction of membrane proteins, such as receptors and cell adhesion molecules, with signalling molecules and neighbouring cells.

Cell recognition and adhesion are processes that involve the carbohydrate chains on the lipid bilayer's extracellular surface. These sugars are crucial to immune responses because they act as markers that tell cells apart.

4. Changing Environments and Adapting to Them:

Due to the dynamic fluidity of lipid bilayers, cells are able to respond to fluctuating external conditions. The presence of cholesterol and the specific fatty acid composition of phospholipids determine the degree of fluidity. It is useful for organisms to have a more fluid membrane in colder surroundings, while it is advantageous to have a less fluid membrane in warmer habitats.

In order to adapt to changes in their surroundings, such as temperature or pressure, cells can alter the lipid content of their membranes. This flexibility allows cells to keep their membranes functioning and intact despite environmental changes.

5. Generation of Energy:

Mitochondria are the "powerhouses" of cells and have lipid bilayers in their double membranes. To generate energy via oxidative phosphorylation, these mitochondrial membranes are required. The production of ATP, the cell's primary energy currency, is dependent on the electron transport chain, a sequence of protein complexes located within the inner mitochondrial membrane.

Sentence #6: Synaptic Transmission

The release of neurotransmitters into the synaptic cleft in nerve cells requires the fast fusion of vesicles holding the neurotransmitters with the presynaptic membrane. Synaptic vesicles rely on lipid membranes for their structure, and lipid membranes mediate the fusion events required for neurotransmitter release.

Immune Reactions:

Cell membrane lipids are involved in immunological responses as well. Pathogen-induced alterations in cell membrane lipid composition can prime the immune system for recognition and response when cells are infected. The detection and removal of dangers to the organism rely heavily on this immunological surveillance.

Finishing Thoughts: Lipid Membranes Serve as the Building Blocks of Life

When it comes to their biological functions, lipid membranes are anything but static limits. They demarcate cellular and organelle boundaries, control molecular traffic, promote cellular communication, and help organisms respond to their surroundings.

The lipid bilayer is the structural basis of all biological life, as it contains both hydrophilic and hydrophobic components. The simple yet adaptable lipid membrane plays a vital part in the orchestration of life's processes, attesting to the elegance and complexity of biological systems. Learning the role of lipid membranes in cellular biology is crucial to comprehending the extraordinary plasticity and efficiency of living things.

Chapter 6.
The proteins are the cellular labourers.

6.1- The chemistry of amino acids and protein structure.

Protein Structure and the Chemistry of Amino Acids

Proteins are the unsung heroes of biology; they power a plethora of critical processes that keep life going. Proteins are incredibly important, but understanding their relevance requires delving into the chemistry of amino acids and learning how the different characteristics and structural arrangements of amino acids create such a wide variety of protein shapes and roles. The chemistry of amino acids and the complexity of protein structure will be dissected in this investigation.

Proteins are composed of amino acids.

Amino acids are the primary structural components of proteins because they are used to construct polypeptide chains. Organic molecules known as amino acids have a core carbon atom (the - carbon) that is bonded to both an amino group (-NH2) and a carboxyl group (-COOH). Each amino acid also possesses a different set of chemical properties thanks to its own side chain, or "R" group.

Structure of Amino Acids:

Group 1 (-NH2) Amino Acids: The amino group is made up of one nitrogen atom and two hydrogen atoms. To create the positively charged ammonium ion (NH3+), it accepts a proton (H+) and hence functions as a base.

Carboxyl (-COOH) Group, Number Two: A carboxyl group consists of a carbon atom that is covalently bonded to two oxygen atoms and a hydrogen atom that is covalently bonded to a single oxygen atom. It has acidic properties, donating a proton to create the carboxylate anion (COO-).

Thirdly, C-Carbon (C-C): The amino acid molecule's core carbon atom provides structural support. It has four different types of bonds: to the amino and carboxyl groups, to hydrogen, and to the R group, which is a special side chain.

The R Group (Side Chain) in 4. Each amino acid has its own unique set of chemical properties, which are a result of the R group. Depending on their chemical make-up, these side chains might be hydrophobic, hydrophilic, acidic, basic, or polar.

Amino Acid Categories:

In living things, protein synthesis is carried out using a set of 20 standard amino acids. These amino acids are distinguished from one another and play varying roles in proteins based on the R group they possess. Some instances are as follows:

To wit: Glycine (Gly) Glycine, which has a hydrogen atom as its R group, is the simplest amino acid. Highly malleable, it contributes structural integrity to proteins like collagen.

This is the amino acid alanine (Ala). Hydrophobicity is due to the methyl group (-CH3) in alanine's R group. It is a staple of proteins' hydrophobic centres.

Note: Serine (Ser): Serine is polar and may form hydrogen bonding because its R group is a hydroxyl (-OH). Enzyme catalysis and protein phosphorylation both depend on it.

Lysine (Lys): Lysine's R group has a lengthy amino group that is positively charged. It has a role in histone modification and in the binding of DNA and RNA.

Peptide bonds are the connecting links between amino acids.

Peptide bonds, which join one amino acid to the next, are essential for protein synthesis. Condensation reaction (dehydration synthesis) between amino group of one amino acid and carboxyl group of another amino acid forms covalent bonds known as peptide bonds. In this process, the -carbons of two amino acids create a bond through the elimination of a water molecule and the formation of a peptide bond.

When two amino acids link together, the resulting structure is called a dipeptide; when three are joined, it's a tripeptide; and so on. It is called a polypeptide when several amino acids are joined in a chain. Proteins are formed when polypeptides fold into their active three-dimensional structures.

The Rank Order of Protein Folding

A protein is made up of a long chain of amino acids linked together by peptide bonds. Proteins, however, do not stay as linear chains; rather, they fold into complex three-dimensional structures. Primary, secondary, tertiary, and quaternary structures are the four tiers of a protein's hierarchical structure.

First, the Basic Structure:

The linear sequence of amino acids in the polypeptide chain is what establishes the protein's main structure. The DNA genetic code specifies this order, with codons corresponding to individual amino acids. The specific arrangement of amino acids in a protein is crucial to the way it works. A change in protein structure and function can result from a single amino acid substitution, and these genetic mutations are commonly the cause of disease.

Secondary framework (2):

Local folding patterns inside a protein are known as its secondary structure. Secondary structures often consist of alpha helices or beta sheets. Hydrogen connections between the backbone atoms (amide hydrogen and carbonyl oxygen) of neighbouring amino acids stabilise these structures.

"Alpha Helix" Hydrogen bonds are formed between the carbonyl oxygen of one amino acid and the amide hydrogen of an amino acid farther along the chain, generating a tightly coiled helical structure called an alpha helix.

Note: Beta Sheet Hydrogen bonds connect adjacent polypeptide chains or segments of the same chain to produce a sheet-like structure known as a beta sheet. The orientation of the polypeptide chains determines whether the beta sheet is parallel or antiparallel.

Tertiary Organisation:

One polypeptide chain's entire three-dimensional folding is called its tertiary structure. Tertiary structure is determined by the precise arrangement of secondary structures (alpha helices and beta sheets) and interactions between amino acid side chains (R groups). Disulfide bridges (covalent connections between cysteine residues) are one type of interaction among many others.

Because it dictates the protein's binding sites and active sites, tertiary structure is fundamental to a protein's activity. It also determines the protein's overall 3D structure.

Fourth, the Quaternary Framework:

When talking about multi-subunit proteins, the quaternary structure describes the organisation of the individual polypeptide chains.

Quaternary structure is unique to proteins with more than two polypeptide chains and is not seen in all proteins. The same interactions that hold tertiary structure together also hold quaternary structure together.

As an example, haemoglobin is a quaternary protein with four distinct tertiary subunits. Haemoglobin is able to efficiently transport oxygen in red blood cells because to the assembly of these subunits.

Protein Function and Folding

Protein function is inextricably tied to the protein's three-dimensional structure, which is established by the protein's primary, secondary, and tertiary structures. The ability of proteins to undergo conformational changes is crucial to their function.

Binding sites in proteins serve as the interface between the protein and its partners, such as substrates and ligands. These interactions between proteins and their binding partners are quite particular because they depend on the proteins' shared structural features and chemical bonds. When a

By binding to a protein, a substrate or ligand can cause the protein to undergo a conformational shift that is crucial to its function.

Proteins, like enzymes, can speed up chemical processes. Where substrates bind and reactions take place, in the active sites, these enzymes have very specific geometries and chemical characteristics. Enzymes often change shape in a way that improves their catalytic activity when a substrate attaches to their active site.

Denaturing Proteins: A Guide to Their Hidden Structures

Proteins' three-dimensional structures are easily disturbed by environmental perturbations such shifts in temperature, pH, or chemical exposure. Proteins are typically rendered ineffective by a process known as denaturation.

The proteins in the egg white (mostly albumin) are denaturized during cooking. When the proteins are heated, they lose their original structure and congeal into a solid mass.

Depending on the degree of structural disturbance, denaturation may be reversible or irreversible. When environmental conditions return to their optimal range, some proteins are able to refold back into their natural form. Some may endure transformations that render them useless forever.

The Chemical Basis for Protein Diversity

Proteins perform a wide variety of biological processes that are necessary for all forms of life. The chemistry of amino acids and the complexity of protein structure are the foundations of their variety and utility.

The incredible complexity of proteins is revealed when their primary, secondary, tertiary, and quaternary structures are understood. Proteins' biological importance stems from their remarkable capacity to fold into highly specific three-dimensional structures, which allows them to recognise and interact with certain molecules and catalyse reactions.

Amino acid and protein structural chemistry not only underpins organismal function, but also provides insights into the molecular basis of disorders, the design of medications, and the investigation of biotechnological breakthroughs. The more we learn about the chemistry of proteins, the more we come to appreciate the beauty and intricacy of life.

6.2- Enzymes and their catalytic role in biology.

Catalytic Enzymes and Their Biological Functions

Enzymes are the conductors of life, weaving together chemical events with incredible speed and precision. These extraordinary biomolecules serve as biological catalysts, hastening processes necessary for cellular and organismal survival. Enzymes play crucial roles in a wide variety of biological processes, and to fully appreciate their importance, it is necessary to investigate their structure, function, and many roles.

How Enzymes Work

As biological catalysts, enzymes reduce the amount of energy normally needed to initiate a chemical reaction. They perform this function without getting depleted themselves, making them not only vital to survival but also highly efficient, useful, and reusable. The role of enzymes in metabolism, energy production, DNA replication, and cell signalling cannot be overstated.

Structure of Enzymes:

Enzymes' structure and function are inextricably intertwined. Generally speaking, enzymes are spherical proteins with a well-defined three-dimensional structure. The enzyme's capacity to interact with the molecule(s) it operates upon depends on its form.

The active site of an enzyme is a specific area whose shape is a perfect match for the substrate(s). An enzyme's active site is the place at which the enzyme binds to its substrate(s). The "lock and key" or "induced fit" paradigm is commonly used to explain enzyme specificity:

Model of a Lock and Key The active site of the enzyme is seen here as a hard structure that fits the substrate snugly, much like a key in a lock.

It's a "Induced Fit Model." In this conception, the active site of the enzyme can conform to the shape of the substrate. The enzyme-substrate fit is improved by this modification, allowing for more efficient catalysis.

The Catalytic and Mechanistic Role of Enzymes

Catalysts, like enzymes, reduce the activation energy required for chemical processes. To commence a chemical reaction, energy must be applied to the reactant molecules in the form of an activation energy. To speed up chemical reactions, enzymes:

Enzymes bring reactants together by positioning substrates such that they can interact more easily. Because of this, less energetic collisions between reactant molecules are required to overcome the activation barrier.

The transition state of a process, an unstable, high-energy intermediate state, is stabilised by enzymes. Enzymes speed up chemical reactions by making the transition state less energetic.

The microenvironment created by an enzyme's active site is optimal for the reaction it catalyses. For instance, if the active site is hydrophobic, it facilitates hydrophobic reactions.

4. Orienting Substrates: Enzymes can orient substrates in a way that facilitates the development of reaction intermediates, thus improving the likelihood of a successful reaction.

An Example of an Enzyme-Substrate Interaction

The complimentary forms and chemical interactions between an enzyme and its substrate(s) make for extremely selective binding. Hydrogen bonds, ionic contacts, hydrophobic interactions, and van der Waals forces are all interactions that enzymes can use to bind substrates. The enzyme-substrate complex is stabilised and the reaction's activation energy is lowered thanks to these interactions.

Nomenclature of Enzymes:

The substrate and catalysed reaction are used to determine the names of enzymes. Each enzyme has been given a unique four-digit identifier called an Enzyme Commission (EC) number as part of a standardised naming system created by the International Union of Biochemistry and Molecular Biology (IUBMB). Hydrolysis of sucrose is catalysed by the enzyme sucrase, which has the EC number 3.2.1.26.

Metabolism-Related Enzymes

Metabolism is the sum total of all chemical events that take place within a cell or organism, and enzymes play a crucial role in this process. Two broad classes of metabolic processes can be identified:

Catabolism, 1. The release of energy occurs during catabolic reactions, which entail the breakdown of complex molecules into simpler ones. Hydrolysis reactions (the addition of water to break bonds) are common in catabolic reactions, hence the enzymes involved in these processes are commonly referred to as hydrolases. Lipases, proteases, and amylases are all enzymes that hydrolyze different types of lipids, proteins, and carbohydrates, respectively.

Two words: anabolism. In an anabolic reaction, energy is used to create a more complex molecule from a simpler one. Condensation or dehydration synthesis reactions, in which water is eliminated to create bonds, are facilitated by enzymes involved in anabolic reactions. DNA polymerases, RNA polymerases, and ribulose-1,5-

bisphosphate carboxylase/oxygenase (RuBisCO) are all enzymes that synthesise certain types of nucleic acids.

Enzyme Activity Regulation

Enzyme activity is strictly regulated to guarantee a steady and effective flow of metabolic processes. Enzyme activity can be adjusted in cells in a number of ways:

First, the Density of the Substrate: When an enzyme catalyses a reaction, the rate of the reaction typically increases as the concentration of the substrate does. The pace of a reaction can be increased by increasing the concentration of the substrate up until the enzyme is saturated.

2. pH: Enzymes have certain pH ranges where they are at their most effective. The enzyme's activity can decrease because of changes in pH, which can break up the ionic connections and hydrogen bonds that keep the enzyme's structure stable.

Thirdly, Temperature: There is a temperature sweet spot where enzymes thrive. Normal reaction speeds increase with rising temperature, however enzymes can be denatured at dangerously high temperatures.

4. Coenz

Enzymes and Cofactors Non-protein molecules called coenzymes or cofactors aid the catalytic activity of many enzymes. While vitamins and other organic molecules serve as coenzymes, inorganic ions like magnesium and zinc can serve as cofactors.

Allosteric Regulation, Number Five: Some enzymes are controlled by allosteric compounds, which attach to allosteric sites and alter the active site's conformation. This may stimulate the enzyme's activity or stifle it.

Sixth, Inhibition of Competition and Other Business Activities: Molecules that bind to an allosteric site, changing the enzyme's conformation, suppress its activity in a manner that is distinct from competitive inhibition, which occurs when two molecules bind to the active site.

Restricting Feedback: The metabolic pathway's final product can reduce an enzyme's activity by blocking its allosteric site. This type of inhibitory feedback is useful for controlling the pathway's overall pace.

The Role of Enzymes in Illness and Healing.

Disrupting the activity of enzymes can have serious consequences for health because they are essential for cell and organism function. Some people are lactose intolerant or suffer from phenylketonuria (PKU) because they lack the enzymes necessary to break down the sugar in milk.

However, faulty enzymes or excessive enzyme activity can also lead to disease. The enzymes that promote erratic cell development may be overproduced by cancer cells. Medicine uses enzyme inhibitors, such as medications and therapies, to block the activity of enzymes thought to play a role in certain diseases.

Summing Up: Life's Catalysts

Enzymes are the unheralded conductors of life's myriad chemical orchestrations. The elegance of biological processes is exemplified by the capacity of enzymes to catalyse reactions with high selectivity and efficiency.

As well as being fundamental to our overall biological knowledge, research into the chemistry and function of enzymes has far-reaching ramifications in domains including medicine, biotechnology, and pharmacology. The more we learn about enzymes, the more possibilities open up for improving human health and expanding our knowledge of the natural world.

Chapter 7.
Nucleic Acids, the Carrier Particles of Information:

7.1- DNA and RNA: their structure and role in genetic information.

DNA and RNA: How They Work to Carry and Store Our Genetic Data

DNA and RNA are the molecular threads in the complicated tapestry of life that carry the instructions for development, function, and heredity. Despite their similarities, these nucleic acids do not perform identical functions in the storage, transmission, or expression of genetic information. The roles of DNA and RNA as the carriers of life's blueprint will be dissected in this investigation of their composition and operation.

DNA is the most important molecule in inheritance.

In most organisms, genetic information is stored mostly in a molecule known as DNA (Deoxyribonucleic Acid). Proteins are the molecular workhorses responsible for almost all biological operations, and DNA contains the instructions necessary to synthesise them.

Genomic structure:

"Double Helix" 1. The double-helix structure of DNA is easily recognised, as it is reminiscent of a twisted ladder or spiral staircase. DNA consists of two very long strands that coil around one other to form a helical structure. In 1953, James Watson and Francis Crick were the first to discover the DNA double helix structure.

Nucleotides are the building blocks of the DNA molecule and are made up of smaller units named adenines and thymine. There are three parts to every nucleotide:
 - Deoxyribose Sugar, also known as a five-carbon sugar molecule.

The deoxyribose sugar has a phosphate group connected to it.
- Nitrogenous Base: Any one of the four nitrogenous bases (A, T, C, and G).

Base Pairing, Step 3 Complementary base pairing is responsible for maintaining the integrity of the two DNA strands. The bases adenine (A) and thymine (T) and cytosine (C) and guanine (G) always pair with one another. Hydrogen bonds keep these base pairs together.

Antiparallel filaments, as in [4.] One strand of DNA is read from 5' to 3', while the other reads from 3' to 5'; these directions are 180 degrees apart. Consistent base pairing is ensured by the antiparallel layout.

DNA's many uses:

1. Data Securing DNA is the molecule that keeps your family's genetic history in its coded form. Cellular machinery interprets this code to synthesise proteins and perform other cellular operations.

Second, a Repetition: Through a process known as DNA replication, DNA may make carbon-copy duplicates of itself. During cell division, this replication process guarantees that genetic material is properly passed on to progeny cells.

Thirdly, DNA is passed on from parent to child during reproduction, a process known as "transmission of genetic information."

DNA stores the information (genes) necessary to create proteins. A similar molecule called RNA is used as a blueprint for protein production after these instructions have been translated from DNA.

RNA: More Than Just a Messenger

RNA (Ribonucleic Acid) is a multifunctional nucleic acid involved in several steps of the production of genes. RNA, in contrast to DNA, may take on multiple distinct structures, each of which performs a particular function.

The RNA Structure:

Nucleotides, first. The nucleotides that make up RNA are structurally similar to those that make up DNA; they consist of a ribose sugar, a phosphate group, and one of four nitrogenous bases. However, uracil (U) is used in place of thymine (T) in RNA. Adenine (A), uracil (U), cytosine (C), and guanine (G) are the four bases found in RNA.

Single-Stranded (2): Base pairing and complementary areas inside an RNA molecule allow for the molecule to fold into intricate three-dimensional structures, despite the fact that most RNA molecules are single-stranded.

RNA Varieties:

There are several different classes of RNA, each of which plays a specific role in various biological operations:

First, there is messenger RNA (mRNA), which transfers genetic instructions from DNA to the ribosomes. Each codon of mRNA bases specifies an amino acid, and the mRNA itself acts as a template for protein production.

Second, transfer RNA (tRNA) molecules are what carry out the translation of the genetic code on mRNA into an amino acid sequence. There is an anticodon area on each tRNA molecule that matches with the corresponding codon in the messenger RNA.

3 rRNA (ribosomal RNA) The protein-making structures of cells called ribosomes rely heavily on rRNA. Protein and ribosomal RNA (rRNA)

work together in ribosomes to catalyse the creation of peptide bonds between amino acids.

Small Nuclear RNA (snRNA): snRNAs function in RNA splicing, the elimination of introns from pre-mRNA to produce mature mRNA.

MicroRNA (miRNA) and Small Interfering RNA (siRNA): By binding to messenger RNA (mRNA) and either blocking translation or increasing the rate of mRNA decay, these tiny RNA molecules serve important roles in controlling gene expression.

The Roles of RNA:

One of RNA's roles is in transcription, the process by which DNA is converted into messenger RNA. Transcription involves the synthesis of an RNA molecule that is complementary to a strand of DNA that serves as a template.

Second, Translation: RNA plays a crucial role in converting messenger RNA (mRNA) into the amino acid sequence that forms proteins. In this process, mRNA, transfer RNA, and ribosomes all work together.

3. Gene Expression Regulation MicroRNAs (miRNAs) and small interfering RNAs (siRNAs) are two types of RNA molecules that regulate gene expression by altering the stability and translation of messenger RNA (mRNA). This regulation is crucial for modulating gene activity in response to external stimuli and during development.

DNA and RNA: The Central Dogma of Genetic Information Flow

The Central Dogma of Molecular Biology outlines the movement of genetic information within a cell in terms of the interaction between DNA and RNA:

Prior to cell division, DNA is copied in order to create new copies of the genetic material.

2. Trans

cription: The process of transcription involves the duplication of a segment of DNA into an mRNA molecule. This mRNA is then shuttled from the nucleus to the cytoplasm to function as a temporary replica of the genetic material.

Translated from 3 Ribosomes and tRNA molecules in the cytoplasm transcribe mRNA into a chain of amino acids. Proteins are built from these particular amino acid sequences.

RNA in Viruses: A Rare Case in Genetics

It's important to remember that not all genes follow the standard DNA-RNA-protein hierarchy. RNA is used primarily by the genomes of certain viruses (termed RNA viruses). These viruses can multiply their RNA genomes without ever touching DNA. The human immunodeficiency virus (HIV) and the influenza virus are both RNA viruses.

The Genetic Symphony, in Conclusion

Gene expression is like a symphony conducted by DNA and RNA, the molecular symphony conductors of life. RNA has dynamic roles in transcription, translation, and gene control, while DNA is the stable storage of genetic instructions.

DNA and RNA work hand in hand to determine an organism's characteristics and capabilities from generation to generation. As our knowledge of their make-up and purpose grows, we get closer to deciphering the mysteries of heredity, evolution, and the complexities of life itself.

7.2- DNA replication, transcription, and translation.

The Molecular Foundations of Life: DNA and Its Replication, Transcription, and Translation

The molecular mechanism that regulates the dissemination of genetic information is an engineering triumph. All living things rely on DNA replication, transcription, and translation to pass on their genetic information and produce proteins, respectively. These fundamental cellular processes determine the characteristics and capabilities of every living organism. To discover the molecular foundations of life, we will take a deep dive into the processes of DNA replication, transcription, and translation.

DNA Replication: A Carbon Copy of the Plan

In order to guarantee that each daughter cell receives an identical set of genetic instructions during cell division, cells must undergo a process called DNA replication. This is the fundamental mechanism through which genes are passed down from one generation to the next.

DNA Replication Catalysts:

To begin with, DNA replication initiates at what are called origins of replication. Here, enzymes known as helicases unwind the DNA double helix into two separate strands.

DNA polymerases are enzymes that catalyse the synthesis of new DNA strands that are complementary to the existing DNA strands, a process known as elongation. The complementary strand is synthesised by using the original strand as a template. Since DNA polymerases can only add nucleotides to the 3' end of a developing strand, this results in the production of two new DNA strands: the leading strand, which is synthesised continuously in the 5' to 3' direction, and the lagging strand, which is synthesised in short segments called Okazaki fragments.

Replication of the DNA molecule stops when the entire molecule has been copied. During cell division, the original DNA molecule splits in two to create two copies, called daughter strands, which are identical to the original.

The Genetic Information Carrier: A Transcript

The process by which a complementary RNA molecule is synthesised from the genetic information contained in DNA is called transcription. Messenger RNA (mRNA) is an RNA molecule that transports genetic instructions from the nucleus to the cytoplasm, where proteins are synthesised.

Procedure for Transcribing:

1. Initiation: The promoter is the starting point for transcription on the DNA molecule. Transcription begins when an enzyme called RNA polymerase attaches to a promoter and opens a gap in the DNA.

During elongation, RNA polymerase acts as a catalyst to produce an RNA molecule by joining together complementary RNA nucleotides. Hydrogen bonds are formed between the RNA nucleotides and the complementary DNA nucleotides as the RNA molecule travels along the DNA template strand. The RNA molecule extends from 5' to 3', following the sequence of the DNA strand serving as a template.

Thirdly, Finishing: When RNA polymerase reaches an endogenous DNA sequence, sometimes known as a "terminator," transcription stops. Here, the newly synthesised RNA molecule and RNA polymerase separate from the DNA template. The genetic instructions are carried by the RNA molecule, now known as the mRNA, which travels to the ribosomes to be translated.

Protein Synthesis is the literal translation.

The translation process is the complex mechanism by which a protein is synthesised from the genetic information contained in messenger RNA. The molecular factories for protein synthesis, called ribosomes, are located in the cytoplasm and are responsible for this intricate process.

Translation Procedures

*First Steps: * The process of translation starts with the tiny ribosomal subunit attaching to the mRNA molecule. In order to begin translating the mRNA into the amino acid methionine (or formylmethionine in bacteria), the ribosome looks for the start codon (AUG). When the small ribosomal subunit binds to the large ribosomal subunit, a functioning ribosome is formed.

Elaboration 2: The ribosome reads codons (three-nucleotide sequences) and recruits transfer RNA (tRNA) molecules as it progresses from the 5' to 3' end of the mRNA. The amino acid carried by each tRNA molecule is the one that corresponds to the codon on the mRNA. Through a process of complementary base pairing, the tRNA molecules are able to connect to the mRNA and bring the amino acids they carry closer together. An expanding polypeptide chain is formed as the ribosome catalyses the creation of peptide bonds between neighbouring amino acids.

Termination, Step 3: Translation proceeds until a stop codon (UAA, UAG, or UGA) is encountered on the messenger RNA. In addition to signalling the end of protein synthesis, these codons do not code for any amino acids. When a stop codon is encountered, the ribosome releases the freshly synthesised polypeptide chain by binding to a release factor protein.

The Genetic Code Is the Language of Life Itself.

The universal language that directs the conversion of nucleotide sequences in messenger RNA into amino acid sequences in proteins is called the genetic code. Each codon on the mRNA is a distinct trinucleotide sequence that specifies one amino acid. Each of the 20 amino acids in proteins has its own unique codon, and there are also three stop codons that mark the end of protein synthesis.

The Molecular Biology Central Dogma

The Central Dogma of Molecular Biology summarises the transmission of genetic information between DNA, RNA, and proteins.

inner to a cell:

Prior to cell division, DNA is copied in order to create new copies of the genetic material.

Transcript 2. The process of transcription involves the duplication of a segment of DNA into an mRNA molecule. This mRNA is then shuttled from the nucleus to the cytoplasm to function as a temporary replica of the genetic material.

Translated from 3 Ribosomes and tRNA molecules in the cytoplasm convert mRNA into the amino acid sequence that proteins require.

The Amazing Dance of Life

Steps in the astonishing dance of life are the replication, transcription, and translation of DNA. These mechanisms underpin our knowledge of genetics, molecular biology, and the translation of DNA into the proteins that control the structure and function of all living things.

Learning more about these molecular processes helps us understand not just how life works, but also how we might make significant advances in sectors like medicine, biotechnology, and genetic engineering. The molecular ballet of life keeps researchers and scientists motivated to unlock the potential of the living world and learn its mysteries.

Chapter 8.
Chemical Reactions Within Cells, or Metabolism

8.1- Exploring the chemical reactions that occur within living cells.

Investigating the Biochemical Processes of Living Cells

Inside every living cell is a symphony of complex chemical events, all working together to ensure the continued existence, development, and reproduction of the organism. In order for cells to obtain energy, construct and repair structures, and perform their specific functions, these reactions must take place. In this investigation, we will delve into the fascinating realm of cellular chemistry and learn about the essential reactions that take place in all living organisms.

Cellular metabolism explains life's chemistry.

Cellular metabolism, the sum total of a cell's chemical reactions, is at the centre of cellular chemistry. There are two primary types of cellular metabolism:

The First Step in the Anabolic Process These biosynthetic reactions are responsible for assembling more complex compounds from their constituent parts. Protein synthesis and the creation of DNA and RNA are examples of anabolic reactions, which need the addition of energy.

Catabolism, secondly: These are the degradative reactions that liberate energy as they reduce complicated compounds to their constituent parts. For example, cellular respiration's breakdown of glucose is a catabolic response.

These two elements of metabolism are inseparably intertwined, establishing a dynamic equilibrium that is essential for cells to keep their structure and function while also extracting energy from their surroundings.

Cellular respiration provides energy for cellular processes.

cellular respiration, the breakdown of glucose to produce energy in the form of adenosine triphosphate (ATP), is a major catabolic activity in cells. There are three major steps in cellular respiration, and they are glycolysis, the citric acid cycle (Krebs cycle), and oxidative phosphorylation (electron transport chain).

The First Step Is Glycolysis

Glycolysis occurs in the cytoplasm of all cells and is an essential metabolic pathway. Glycolysis is the metabolic process through which glucose, a six-carbon sugar, is broken down into pyruvate, a three-carbon substance. A total of four ATP molecules and two molecules of the high-energy electron transporter NADH are produced from just two ATP molecules.

The Citric Acid Cycle: Central to the Molecular Machine

The mitochondria then transport the pyruvate molecules to the citric acid cycle, where they are oxidised further. Each pyruvate undergoes a carbon dioxide reduction and an electron transport chain reaction, producing two molecules each of NADH and $FADH_2$. In cellular respiration, the citric acid cycle serves as a hub that connects and generates high-energy molecules for subsequent steps in the process.

Redox phosphorylation: the power plant.

The majority of ATP is generated in the electron transport chain, which is located within the inner mitochondrial membrane during cellular respiration. Energy is generated as electrons from NADH and

FADH2 are shuttled through a chain of protein complexes. To generate an electrochemical gradient, mitochondria utilise this energy to pump protons (H+ ions) across their membrane.

ATP is synthesised from adenosine diphosphate (ADP) and inorganic phosphate (Pi) by the re-entry of protons into the mitochondrial matrix via a protein complex termed ATP synthase. The majority of the ATP used by the cell is produced by this pathway, which is called oxidative phosphorylation.

The Anaerobic Process of Fermentation

fermentation, an anaerobic process, allowing glycolysis to proceed in oxygen-starved cells. Pyruvate is fermented in animals to produce lactic acid, and in yeast and some bacteria to produce ethanol and carbon dioxide. Cells can generate a little amount of ATP and renew NAD+ by fermentation, which keeps glycolysis going despite its lower ATP output compared to aerobic respiration.

Green energy production via photosynthesis.

Amazingly, plants, algae, and even certain bacteria are able to transform the energy from sunlight into chemical energy stored in glucose and other organic molecules through a process called photosynthesis. Light-dependent processes and light-independent reactions (the Calvin cycle) make up the two halves of photosynthesis.

Reactions in Response to Light: Soaking up the Sun

The light-dependent reactions involve the absorption of light by chlorophyll and other pigments in the thylakoid membranes of chloroplasts. High-energy molecules like ATP and NADPH are produced, and oxygen is released as a consequence, from this process.

In the absence of light, glucose is synthesised via the Calvin cycle.

Light-dependent processes produce ATP and NADPH, which are then used in the Calvin cycle in the chloroplast stroma to convert carbon dioxide into glucose and other carbohydrates. Carbon dioxide is reduced to carbon monoxide and ATP and NADPH are produced by a sequence of enzyme events in the Calvin cycle.

For all heterotrophic species that rely on ingesting plants or other photosynthetic organisms, photosynthesis serves as the cornerstone of the food chain, providing them with energy.

The Role of DNA Repair in Preserving Genetic Purity

Chemical agents, radiation, and replication mistakes are all potential threats to the DNA in a cell. Cells use a variety of DNA repair mechanisms to ensure the stability of their genetic material.

Nucleotide excision repair is one such mechanism; it locates and eliminates harmed DNA sequences. Several enzymes work together to identify the damaged area, remove the offending segment of DNA, and insert the proper sequence in its place.

Mismatch repair, which fixes mistakes made in DNA replication, is another crucial mechanism. Mismatched bases that were accidentally integrated into the newly synthesised DNA strand are detected and removed by this technique.

DNA repair systems are crucial for preserving genetic information and halting the development of cancer and other disorders caused by mutations.

The Molecular Workers of Enzymatic Catalysis

Catalysts are molecules, and enzymes are the molecular labourers behind

the plethora of cellular chemical processes. In order for processes to take place at temperatures and in timescales relevant to living organisms, these specialised proteins reduce the activation energy necessary for them.

The processes that each enzyme catalyses are quite specific to that enzyme. Substrates (reactant molecules) bind to an enzyme's active site, where the substrates undergo a chemical change and the products are released. The precise arrangement of amino acids in enzyme active sites is what gives them their specificity.

Coordinated reactions along metabolic pathways

A metabolic route is a network of interconnected enzyme processes that work together to provide a certain biological function. Each reaction along these pathways is catalysed by a different enzyme, and the pathways themselves can be either linear or branched.

Earlier, we discussed glycolysis in the context of cellular respiration as an example of a well-known metabolic route. The breakdown of glucose into pyruvate and ATP and NADH occurs through a series of ten enzyme processes known as glycolysis.

The Chemical Language of Cells

Through a system of cell signalling pathways, cells are able to interact with one another and adapt to their surroundings. In these processes, ligands are bound to cell surface receptors to initiate a chain reaction inside the cell.

Insulin signalling, for instance, is essential in controlling blood sugar levels. The pancreas secretes insulin in response to a rise in blood glucose after a meal, and this insulin attaches to insulin receptors on the surface of specific cells. This causes a cascade of biological events that ends with glucose being taken up from the blood and stored as energy by the cells.

The Chemical Symphony of Life, Final Thoughts

Like the notes and melodies of a symphony, the chemical reactions within living cells work together in harmony to keep the organism alive. Each part of this complex symphony—cellular metabolism, photosynthesis, DNA repair, enzymatic catalysis, metabolic pathways, and cell signaling—is essential to the health of the cell and the organism it is part of.

The chemical basis of life is becoming increasingly clear as scientists continue to decipher the mysteries of cellular chemistry. This information not only deepens our comprehension of biology, but also promises to pave the way for advances in medicine, biotechnology, and the creation of long-term answers to global problems. Exploration and discovery are continually sparked by life's molecular symphony, which is gradually revealing its secrets.

8.2- The role of enzymes in metabolic pathways.

Enzymes, Nature's Molecular Catalysts, and Their Function in Metabolic Pathways.

Enzymes are the unsung heroes that power the chemistry of life within the complex web of cellular activities. These extraordinary molecules work as molecular catalysts, speeding up metabolically crucial chemical events that are vital to life's delicate balance. Through this investigation, we will learn how enzymes are used in metabolism and the incredible efficiency and specificity with which they function.

Molecular labourers in nature, enzymes.

The activation energy needed for a chemical reaction to take place is lowered by enzymes, which are biological macromolecules that are typically proteins. They facilitate and regulate a wide variety of chemical reactions within cells, making them the genuine workhorses of life. Many of life's fundamental functions would grind to a halt without enzymes, making it impossible to maintain the cellular activity necessary for survival.

Catalysis: The Advantage of Enzymes

Catalysis, the acceleration of a chemical reaction without the catalyst's own depletion or alteration, is crucial to understanding how enzymes work. To facilitate the conversion of reactants into products, enzymes create a new reaction pathway with a lower activation energy.

To overcome the activation energy barrier, which is the lowest amount of energy needed to get molecules from the reactant state to the transition state (where new bonds can be formed) is a major challenge for many chemical reactions. Enzymes catalyse reactions by stabilising the transition state, which lowers the activation energy and speeds up the reaction.

The Lock and Key of Enzyme-Substrate Interaction

The efficiency of enzymes relies on the fact that they work in a very specific way. The substrate (the molecule on which the enzyme operates) binds in a specific pocket or crevice (the active site) in the three-dimensional structure of each enzyme. The substrate acts as the key that fits perfectly into the active site of the enzyme, which is sometimes compared to a lock and key mechanism.

The precise arrangement of amino acid residues in the active site produces a complimentary environment for the substrate, resulting in the high specificity of enzyme-substrate interactions. This selectivity guarantees that enzymes catalyse just the reactions for which they were evolved, limiting the possibility of unintended cellular reactions.

The Lock and Key Mechanism of Enzyme Catalysis

Several steps make up an enzyme's catalytic cycle:

Binding to Substrates 1. The substrate reaches the enzyme's active site, where it binds. The enzyme-substrate complex is formed when the substrate binds tightly to the active site, a process that requires a high degree of specificity.

Second, Stabilising the Transitional State: The activation energy of a reaction is lowered after the enzyme has been bound because the transition state is stabilised. Interactions like as hydrogen bonding, electrostatic repulsion, and form complementarity all play a role in this stabilisation.

Third, Product Development: The enzyme helps speed up the reaction that produces the product from the substrate. Without the enzyme, the process would be slower and the transition state would be achieved less quickly.

4. Release of Product The enzyme then releases the product and its active site is free to interact with a new substrate molecule, allowing the reaction to repeat.

5. Recycling of Enzymes Importantly, enzymes are stable enough to catalyse the same process several times without degrading.

Fine-Tuning Cellular Activities via Enzyme Regulation

By modulating enzyme activity, cells are able to keep tight rein on their metabolic processes. There are a number of ways that enzyme activity can be altered.

Some enzymes have what are called allosteric sites, which function independently of the active site. Enzyme activity and conformation can be modulated by molecules attaching to specific locations on the enzyme.

In competitive inhibition, molecules with a similar structure to the substrate battle it out for access to the active site. Substrate binding is inhibited when an inhibitor is present in the active site.

The third type of inhibition is called "non-competitive inhibition," and it occurs when an inhibitor binds to a part of an enzyme other than the active site. Because of the conformational change brought about by this interaction, the enzyme's catalytic activity is diminished even in the presence of its substrate.

Covalent modification (fourth): Covalent modifications, such as phosphorylation, can be applied to some enzymes to either stimulate or suppress their activity. Phosphate groups are added by protein kinases and removed by phosphatases.

Inhibition from Feedback (Point 5) The end product of a metabolic pathway can sometimes inhibit an enzyme further up the system via feedback inhibition. This control system keeps the levels of intermediate metabolites stable and stops the body from making too much of the final product.

Metabolic Pathways as Illustrations of Enzyme Action

In numerous metabolic pathways, enzymes play crucial roles in maintaining life. Some instances are as follows:

Glycolysis, 1. Glucose is converted into pyruvate through a sequence of enzyme events in the central metabolic pathway, which also results in the production of ATP and NADH. Vital enzymes in the glycolytic pathway include hexokinase, phosphofructokinase, and pyruvate kinase.

Two, the Krebs Cycle (

Citric Acid Synthesis): Citrate synthase, isocitrate dehydrogenase, and -ketoglutarate dehydrogenase are only a few of the enzymes involved in the Krebs cycle in the mitochondria. To generate NADH and FADH2 for the electron transport chain, these enzymes catalyse the oxidation of acetyl-CoA.

Replicating DNA, or Thirdly: DNA replication requires enzymes called DNA polymerases to catalyse the production of new DNA strands. They ensure that genetic information is replicated accurately by adding nucleotides that are complementary to the template strand.

During the Calvin cycle, enzymes like ribulose-1,5-bisphosphate carboxylase/oxygenase (RuBisCO) fix carbon dioxide, which results in the formation of glucose and other organic compounds during photosynthesis.

Enzymes as a Resource in Biotechnology

Because of their exceptional catalytic characteristics, enzymes have found widespread use in both biotechnology and industry. Among the many uses they have are:

Amplification of specific DNA sequences for research and diagnostics is made possible by the polymerase chain reaction (PCR), which uses DNA polymerases to repeat DNA segments.

2. Bioprocessing: Enzymes are used in the manufacturing of food, beer, and cleaning products to aid the breakdown of complicated substrates.

Bioremediation, in which enzymes are utilised to eliminate pollution and decompose environmental toxins, is a viable option for long-term environmental purification.

Pharmaceuticals, such as antibiotics, hormones, and therapeutic proteins, rely on enzymes in their manufacture.

Finally, enzymes are Nature's own catalysts.

Molecular catalysts known as enzymes are responsible for catalysing a wide variety of biochemical reactions. They are crucial to metabolic pathways and various biological functions due to their specialisation, control, and adaptability. Discoveries on the mechanisms that control enzyme activity pave the way for new biotechnological applications and shed light on the chemical orchestra that is life. Catalysts in nature, enzymes are responsible for the chemistry of life and hold the key to a better future in biotechnology and medicine.

Chapter 9.
Signalling molecules and hormones:

9.1- The chemistry behind signaling molecules and their impact on cellular communication.

The Role of Signalling Molecules in Cellular Communication and Their Chemistry

Communication is crucial in the complex biological landscape. To keep the delicate balance of life, cells must communicate, react to their surroundings, and work together. Small and big signalling molecules are the messengers that carry messages within and between cells. A fascinating tale of specialisation, diversity, and precision unfolds in the chemistry behind these molecules and their interactions. Through this investigation, we will learn more about the chemistry of signalling molecules and the role they play in cellular communication.

Communication via cell phone: a necessity for survival.

Everything from tiny single-celled organisms to large multicellular animals like humans relies on cellular communication to function. Fundamentally, cellular communication is the sharing of information between cells or within a cell. By communicating with one another, cells are able to adapt to their surroundings, work together within a tissue, and control vital functions including growth, development, and immunity.

The Signalling Molecules Are Life's Transmitters

signalling molecules constitute the backbone of cellular communication, relaying messages from one cell to another. There are two main classes into which these molecules fall:

1. Signalling molecules (or peptides) Typically, these are tiny molecules with a high rate of action due to their ability to diffuse through cell membranes. Neurotransmitters, hormones, and gasotransmitters like nitric oxide (NO) are all good examples.

Big molecules that send messages Proteins and nucleic acids are examples of such macromolecules, and they normally trigger responses via specialised cell surface receptors. Some examples are cytokines, growth factors, and DNA.

Chemical messengers, or small signalling molecules

When it comes to intercellular communication, little signalling molecules are crucial. Let's take a look at the chemistry and mechanics that make these transmitters tick:

Hormones: Hormones are signalling molecules that glands secrete into the bloodstream. From there, they can go to their intended recipient cells. Insulin, produced by the pancreas, and adrenaline, produced by the adrenal glands, are two examples of the hormones used by the endocrine system to control metabolic processes and the body's response to stress.

Neurotransmitters: The tiny molecules known as neurotransmitters play an essential role in the communication process between neurons and other cell types. Serotonin, dopamine, and acetylcholine are all neurotransmitters. Neurons produce these molecules, pack them away in vesicles, and then secrete them into synapses, where they can bind to their respective receptors on target cells.

Gasotransmitters: Gaseous mediators of signalling include nitric oxide (NO), carbon monoxide (CO), and hydrogen sulphide (H2S). For instance, endothelial cells produce nitric oxide, a vasodilator that widens blood arteries to control blood pressure.

Proteins and nucleic acids are examples of large signalling molecules.

Proteins and nucleic acids, two types of large signalling molecules, play important roles in a variety of biological activities. We'll delve into their chemical make-up and function in intercellular communication below.

Proteins: Proteins involved in signalling have many different roles. Some, like receptor proteins, function as gatekeepers, transmitting information from the extracellular environment into the cell. Others, like kinases, modify proteins in specific ways by adding phosphate groups (phosphorylation). In the process of signal amplification, in which a single extracellular signal can set off a series of intracellular actions, proteins also play important roles.

Hormonal Peptides: Peptide hormones, which are common, are made up of relatively small chains of amino acids. Insulin, a peptide hormone, is one such example; it controls glucose levels in the blood. Precursor molecules for peptide hormones are synthesised and then cleaved by enzymes to yield the functional hormone.

Cytokines: Signalling proteins called cytokines control immunological responses, inflammation, and cell proliferation. Both in health and illness, they are crucial players. Cytokines include molecules like interleukins, interferons, and tumour necrosis factor (TNF).

The Role of DNA as a Message Carrier: DNA itself can act as a messenger. DNA strand breaks or lesions can trigger the activation of repair mechanisms and the stimulation of cell cycle checkpoints in DNA damage response pathways.

Specificity at the Molecular Level of Chemical Receptors

Signalling molecules must bind to their respective receptors on the surface of the cells they are trying to communicate with, or within the cytoplasm or nucleus of the cells themselves. These connections are so precise that they are sometimes equated to a "lock and key" system, with the signalling molecule serving as the key and the receptor serving as the lock.

This exquisite precision is the result of amazing chemistry. Each receptor recognises a unique signalling chemical, and its binding site is shaped and charged specifically to attach to that molecule. Because of their perfect fit together, the receptor can be activated by only the intended signalling molecule. If a molecule isn't a perfect match for the binding site on a receptor, not even a very similar one will bind to it.

Transmission of Messages; Signal Conversion

When a signalling molecule interacts to its receptor, it triggers a chain reaction of biological processes. This is accomplished through a series of chemical reactions known as signal transduction, which serves to both amplify the signal and convey it into the cell's interior. Messages from receptors are transmitted to effector proteins with the help of second messengers like cyclic AMP (cAMP) and calcium ions (Ca^{2+}).

Examples of Pathways in Cellular Communication

Real-world examples are crucial for grasping the significance of signalling molecules and the chemistry behind cellular communication.

Signalling via Insulin: 1. When glucose levels in the blood become too high, insulin is secreted by the pancreas. The binding of insulin to cell surface receptors initiates a signalling cascade that promotes glucose uptake from the blood into the cell. The result is a normalisation of blood sugar.

Second, "neurotransmission" Dopamine and other neurotransmitters are essential for proper brain function. Neurotransmitters are chemical messengers that are released into synapses, where they interact with postsynaptic receptors.

neurons, which send out messages that influence one's disposition, actions, and thoughts.

The Immune System: In reaction to an infection or tissue damage, immune cells release cytokines like interleukin-1 (IL-1). They attach to receptors on immune cells, setting off a signalling cascade that drives inflammation and brings in more defence cells.

Growth factors, such as epidermal growth factor (EGF), stimulate cell growth, proliferation, and differentiation by binding to receptors on cell surfaces and sending out a signal. Diseases like cancer can develop when growth factor signalling goes awry.

Chemical Variation in Transmitting Messages

The chemical structures and characteristics of signalling molecules span a wide range. The neurotransmitter serotonin is an example of a very straightforward signalling molecule, while the protein hormone insulin is an example of a relatively complex one. Different chemical structures serve different purposes and have different levels of specificity, both of which are necessary for efficient cellular communication.

Summing Up: Translating Life's Chemical Code

Understanding the chemistry of signalling molecules and how it affects cellular communication is an intriguing glimpse into the workings of the living world. These large and small molecules convey instructions that direct cellular activity and conduct the orchestra of life. The possibility of new medicines and treatments for a variety of diseases and ailments rests on our ability to decipher their chemistry and mechanisms, which in turn expands our understanding of biology. Scientists are still trying to understand the secrets encoded in life's molecular language.

9.2- Examples of hormone signaling in the body.

Hormone Signalling Examples: Conducting a Symphony of Optimal Health and Balance.

Hormones serve as chemical messengers that regulate a wide variety of bodily functions. Amazingly, these chemicals control not only development and metabolism, but also emotions and the immune system. We will examine some important examples of hormone signalling in the body, focusing on the critical functions they play in preserving health and equilibrium.

Hormones serve as the "messengers" of the body.

Signalling molecules called hormones are made by the endocrine system's specialised glands, tissues, or cells. Once released, they can circulate throughout the body and act on specific cells wherever they happen to be. The harmonious functioning of the body depends on hormones' ability to regulate a wide variety of physiological activities.

First, Insulin Controls Blood Glucose Levels

The pancreas secretes a number of hormones, but one of the most well-known is insulin. Blood glucose regulation is a crucial part of energy metabolism, and insulin plays a major role in this process. Blood glucose levels increase after eating carbohydrates. As a result, insulin is secreted into the bloodstream by the pancreas.

Cells can only take in glucose from the bloodstream if insulin first opens their doors. Glucose is converted into immediate energy in the mitochondria or stored as glycogen in the liver and muscles. This method lowers blood sugar levels and keeps energy flowing to cells at all times.

Thyroid hormones are responsible for regulating metabolic rate.

The thyroid gland secretes a variety of hormones, including thyroxine (T4) and triiodothyronine (T3). They play a crucial role in metabolic regulation, which in turn affects how quickly or slowly the body transforms food into energy. Heart rate, core temperature, and energy expenditure are just few of the many bodily processes that thyroid hormones influence.

Fatigue, weight gain, and sensitivity to cold are all symptoms of hypothyroidism, a disorder caused by low levels of thyroid hormone. On the other hand, hyperthyroidism can cause symptoms like a racing heart, decreased appetite, and increased sensitivity to heat.

Cortisol, Stress, and Coping

The adrenal glands secrete cortisol, also known as the "stress hormone," in reaction to stressful situations or low blood glucose. It's essential for the body's stress response, where it helps channel energy and regulate the body's response to stress.

The "fight or flight" reaction to stress is characterised by an increase in cortisol levels, which in turn causes an increase in glucose production, blood pressure, and attentiveness. The body is getting ready to react to real or imagined danger. However, increased cortisol levels are associated with chronic stress and have been linked to anxiety, sadness, and the suppression of the immune system.

Oestrogen and progesterone: Controlling Fertility

The ovaries in females and the testes to a lesser extent in males are responsible for the production of the sex hormones oestrogen and progesterone, respectively. These hormones are crucial in controlling the reproductive system of females.

Secondary sexual traits, such as the growth of breast tissue and the onset of menstruation, are influenced by oestrogen. It's important for bone health and for keeping your mood stable, too.

On the other hand, progesterone is essential for priming the uterus for conception and maintaining a healthy pregnancy in its early stages. It plays a role in keeping the uterine lining healthy and in regulating the menstrual cycle.

Testosterone's Power to Shape Men's Identities #5

Males' secondary sexual features, such as facial hair, a deep voice, and muscular growth, are largely attributable to Testosterone, which is principally produced by the testes in men. It's also important for a man's libido and sperm count.

The ovaries create a trace amount of testosterone, which is crucial for women to keep their muscle and bone mass. Infertility, emotional instability, and sexual dysfunction are just some of the problems that have been linked to low testosterone levels.

6. Growth Hormone: Aiding in Regeneration and Maintenance

The pituitary gland's growth hormone (GH) is critically important for promoting growth in young people. It stimulates the creation of insulin-like growth factor-1 (IGF-1), which mediates many of GH's actions, including bone and tissue growth.

In addition to its role in development, growth hormone (GH) also helps keep adult muscles and bones strong and healthy. It helps with tissue repair and general body composition as well. Both gigantism and dwarfism can result from GH production or regulatory problems.

Melatonin, the Hormone of Sleep and Wakefulness 7

Pineal gland in the brain secretes a hormone called melatonin largely at night in response to dim lighting. It is essential in maintaining the circadian rhythm, or the body's natural sleep-wake cycle. The body's production of the sleep hormone melatonin increases as the day winds down, indicating that it is about time to turn in.

Insomnia, as well as jet lag, are two frequent sleep disorders that melatonin supplements are used to treat. Melatonin aids in sleep-wake cycle synchronisation by modulating the circadian clock.

Parathyroid Hormone, Number Eight

Calcium Homeostasis (PTH)

Parathyroid glands in the neck secrete parathyroid hormone (PTH), an important hormone in controlling blood calcium levels. In addition to increasing calcium absorption in the kidneys and intestines, PTH also stimulates the release of calcium from bones into the bloodstream.

Keeping your calcium levels in check is important for your bones, muscles, and nerves. Hyperparathyroidism (excessive PTH production) and hypoparathyroidism (insufficient PTH production) are two diseases that can result from PTH regulatory problems.

Nine. Epinephrine (Adrenaline) and the "Fight or Flight" Physiology

The adrenal glands secrete adrenaline (also known as epinephrine) in reaction to stress or the perception of danger. The "fight or flight" reaction, in which the body gets ready to take action quickly in response to threat, relies heavily on this hormone.

Adrenaline helps the body respond to stress by speeding up the heart rate, widening the airways, and diverting blood flow to the muscles. While this reaction is essential in times of danger, its prolonged use might have negative effects on health.

Leptin, Number Ten: Suppressing Hunger

Leptin is a hormone that aids in controlling hunger and maintaining a healthy weight, and it is secreted by adipocytes (fat cells). It aids in energy regulation by sending signals to the brain when fat stores are full.

An increase in fat mass causes an increase in leptin, which in turn causes a decrease in hunger and an increase in energy expenditure. However, obesity and metabolic diseases can be exacerbated by leptin resistance, which occurs when the brain does not respond normally to leptin.

11. Oxytocin, Also Known as the "Love Hormone"

Some people even refer to Oxytocin as the "cuddle hormone" or the "love hormone." Social bonding, emotional connection, and uterine contractions all depend on this hormone, which is secreted by the pituitary gland after being created in the hypothalamus.

Hugging, cuddling, and breastfeeding all result in the release of oxytocin, which has been shown to improve trust and interpersonal bonds. The stimulation of uterine contractions by this hormone is also crucial during labour and delivery.

Vasopressin (Antidiuretic Hormone): Controlling Fluid Intake and Excretion 12

Vasopressin, or antidiuretic hormone (ADH), is secreted by the pituitary gland after being generated in the hypothalamus. Controlling the kidneys' ability to reabsorb water is essential for maintaining a healthy water balance and blood pressure.

Urine concentration and blood volume both rise with elevated vasopressin because the kidneys reabsorb more water. However, when vasopressin levels are low, urine production increases. Diabetes insipidus and the syndrome of inappropriate antidiuretic hormone (SIADH) are disorders that can arise from vasopressin dysregulation.

The Complex Dance of Hormone Signalling

The intricate and strictly controlled process of hormone signalling in the human body affects virtually every facet of our health and well-being. Numerous physiological functions, including development, metabolism, reproduction, and reactions to stress, are coordinated by chemical messengers. Hormones play crucial roles in health and disease, and understanding their complicated dance inside the body is essential for prevention and treatment. The astonishing complexity of human biology rests on a foundation of hormone signalling, which is a symphony of precision, balance, and adaptation.

Chapter 10.
Health and Chemistry:

10.1- How understanding chemistry can lead to advances in medicine and drug development.

The Contribution of Chemical Knowledge to the Progress of Medicine and Drug Development.

When it comes to medicine and the creation of new drugs, chemistry is the foundational science that underpins our understanding of the molecular world. Chemistry plays a crucial role in improving our capacity to detect, treat, and prevent disease, from deciphering the structures of biomolecules to creating innovative medicinal chemicals. In this investigation, we'll dig into how a better grasp of chemistry has facilitated major advances in healthcare and drug development.

Science Reveals the Building Blocks of Life: Chemistry and Molecular Biology

Molecular biology, the study of biomolecules at the atomic and molecular level to determine their structures and functions, is fundamental to the study of biology and medicine. Understanding the language of life's blueprints requires the tools and concepts chemistry gives.

The discovery of the DNA double helix by James Watson and Francis Crick in 1953 was a seminal moment in the history of biology. Chemical principles applied to the study of X-ray diffraction data allowed for this ground-breaking insight. The discovery of DNA's structure cleared the door for the decoding of the genetic code, the isolation of disease-causing genes, and the creation of effective genetic therapies.

Enzymes and Proteins: Proteins serve as the "workhorses" of biology, as they are responsible for a wide variety of cellular processes. Structures, folding mechanisms, and catalytic activity of proteins are all clarified by the study of protein chemistry. Enzymes are a type of protein that plays a crucial role in almost all biological reactions. Enzyme-based medicines and diagnostics have advanced thanks to research into enzyme kinetics and processes.

3. Drugs Aimed Towards: Disease-causing chemicals can be pinpointed with the use of chemistry. Drugs can alter the function of these molecules, which are typically proteins or nucleic acids. The development of therapeutic medicines with selective binding to specific targets requires an understanding of their chemical characteristics.

Therapeutic Compound Chemistry (Drug Discovery)

The design, synthesis, and testing of new chemicals for their potential as medications typically begins in the laboratory, where chemists and pharmacologists collaborate. Chemical concepts are crucial to the drug discovery process.

1. Methods for Efficient Drug Design The three-dimensional architectures of biological molecules are used in rational drug design to provide medications with targeted effects. Drugs with high specificity and potency can be developed by studying the target's chemistry and developing compounds that fit the binding site.

Two words: High-Throughput Screening To find promising medication candidates, researchers first analyse chemical libraries containing thousands of chemicals. High-throughput screening uses automation and robots to efficiently evaluate huge numbers of chemicals.

Medical Chemistry, Third: Medicinal chemists work to improve the potential therapeutic effects of lead compounds. Medicinal chemists work to enhance the pharmacokinetics, bioavailability, and safety characteristics of lead compounds by altering their chemical structures.

Computer-assisted drug design, or CAD: Researchers can model drug-receptor interactions, estimate binding affinities, and speculate on how new compounds can interact with biological targets by employing computational chemistry and molecular modelling approaches. Because fewer molecules are required for synthesis and testing, the drug development process is sped up.

Here are several Instances of Drug Discoveries Fueled by Chemistry

Chemical ideas and methods have been used to develop many different medicines. Some instances are as follows:

(1) Penicillin: The age of antibiotics began with Alexander Fleming's 1928 discovery of penicillin. Howard Florey and Ernst Boris Chain determined the chemical structure of penicillin, a natural substance extracted from the Penicillium fungus. Because of this breakthrough, bacterial illnesses can finally be effectively treated.

Second, aspirin: Aspirin, one of the most used drugs in the world, was first created from willow bark. Scientists in the late 19th century synthesised and characterised acetylsalicylic acid, the active component. Because of its effectiveness in reducing inflammation and easing pain, aspirin has become a standard treatment for both conditions.

Thirdly, statins: Commonly prescribed medicines for reducing cholesterol and cardiovascular disease risk include statins like atorvastatin and simvastatin. In order to create these medications, researchers had to learn the biochemistry of cholesterol manufacturing and then build chemicals that block the enzyme HMG-CoA reductase.

Monoclonal Antibodies, Number Four Rituximab and trastuzumab are two examples of the type of targeted therapy known as monoclonal antibodies that are used to treat cancer and autoimmune illnesses. Developing antibodies that preferentially bind to disease-specific molecules requires a combination of immunology and molecular biology methods, with an emphasis on precision chemical engineering.

Diagnostic Chemistry: Molecular Markers in Personalised Medicine.

Early disease identification, disease monitoring, and individualised treatment plans are all made possible by diagnostic instruments made possible by chemistry. Tools of this type frequently rely on the detection of biomarkers, chemicals characteristic of a given disease or physiological state:

* 1. Diagnostic Radiology * Imaging methods that provide anatomically accurate pictures of the human body rely on chemical principles. These methods include magnetic resonance imaging (MRI), positron emission tomography (PET), and computed tomography (CT). Tissues and organs can be better seen when contrast agents containing such substances are used.

secondly, Liquid Biopsies The analysis of blood, urine, or other bodily fluids for biomarkers linked with diseases like cancer is known as a liquid biopsy, and it is a non-invasive diagnostic technique. Chemical assays are frequently used in these tests for the detection and quantification of target compounds like circulating tumour DNA or cancer antigens.

Thirdly, Mass Spectrometry: Mass spectrometry is a highly accurate method for identifying and quantifying biomolecules. It can be used in the fields of proteomics, metabolomics, and the study of small molecules, making it a powerful resource for medical research.

Genomic Sequencing, Number Four: DNA sequencing and analysis are essentially chemical processes, even though genomics is a subfield of molecular biology. Extensive chemical procedures were used, for instance, to sequence the whole human genome as part of the Human Genome Project.

Directions for the Future: Where Chemistry Meets Medicine

When chemistry and medicine work together, amazing things can happen in the medical field. Here are a few areas where chemistry is expected to make major strides in the near future:

1. Personalised Medicine: Genomic, proteomic, and metabolomic advancements are paving the way for personalised medicine, in which patients receive care based on their unique genetic and molecular profile. The potential for more precise and less harmful treatments is enhanced by this method.

Systems for Administering Medication Improved drug solubility, stability, and targeted distribution are all possible thanks to chemistry-based drug delivery systems. Drugs are being engineered to be transported to disease areas using nanoparticles, liposomes, and other nanoscale carriers.

Third, Synthetic Biology: To create novel biological systems and species, synthetic biologists apply chemical and biological concepts. New macromolecules, medicinal proteins, and bio-based materials could all be produced with the help of this developing sector.

Work on Vaccines, Number Four: From antigen design and synthesis to adjuvant development and formulation, chemistry plays an essential part in the vaccine development process. The prevention and treatment of new infectious illnesses, as well as the enhancement of vaccine efficacy, depend on developments in vaccine chemistry.

Medicine's Constant Development Due to Chemical Advances

Innovation and improvement in medical care are sustained by the close bond between chemistry and medicine. The application of chemical principles to drug discovery and diagnostics, combined with our molecular biology knowledge, has resulted in revolutionary medical advances. As we learn more about the human body and the diseases that plague it, chemistry will remain a crucial tool for discovering new ways to treat and prevent illness and for expanding our understanding of the intricate chemistry of life. New opportunities for treatment, prevention, and personalised care are waiting to be discovered at the interface of chemistry and biology, which is where medical progress will be made in the future.

10.2- The impact of chemistry on our daily lives and health.

The Effects of Chemistry on Our Everyday Lives and Physical Well-Being

Chemistry is often called the "central science" because of the many ways in which it affects and is influenced by everything we do and consume. Medicine, environmental quality, and consumer product safety all rely on it, therefore its effects on human health are far-reaching. To demonstrate that chemistry is essential to modern life and a key factor in our well-being, we shall delve into its far-reaching influence on our daily lives and health.

Our Chemical Connection to the Food We Eat

The chemistry of food is crucial to our daily lives since it impacts the food's nutritional value, flavour, and safety. Important ways in which chemistry affects the food we eat include:

One, a breakdown of the nutrients: The nutritional value of foods is mostly determined by their chemical composition. Vitamins, minerals, and macronutrients (carbohydrates, proteins, and fats) may all be accurately measured thanks to analytical methods including spectrophotometry, chromatography, and mass spectrometry.

Second, food preservation techniques utilise chemistry to create a longer-lasting product. Methods that rely on regulating chemical reactions and microbiological growth fall within this category, and they include canning, pasteurisation, and refrigeration.

3. Chemical Basis of Flavour Complex chemical reactions occur between chemicals to produce a food's aroma and flavour. The study of flavour chemistry delves into the volatile chemicals that give meals their distinctive scents and flavours.

(4) Food Security: Maintaining safe food supplies relies heavily on chemistry. Consumers can be safeguarded from foodborne illness and allergic reactions with the aid of techniques like food testing, microbiological analysis, and the detection of pollutants (including pesticides, additives, and allergens).

In order to enhance the quality of foods in terms of its texture, flavour, and visual appeal, many different types of food additives have been developed and tested. These additives are chemically guaranteed to be harmless.

The Role of Chemistry in Medicine

Chemical processes are crucial to the discovery and manufacture of new medications. Here is how chemistry influences our health by way of medicine:

Discovering New Drugs The search for and development of novel pharmaceuticals is under the purview of medicinal chemistry. The discovery of new therapeutics relies heavily on the application of chemical principles, including target identification, compound screening, and lead compound optimisation.

Formulating medications into different dose forms like tablets, capsules, and injections is an important part of pharmaceutical chemistry. These compositions' stability and bioavailability are guaranteed by chemical processes.

3: Pharmacokinetics Determining dose schedules and optimising medication therapy require an understanding of how pharmaceuticals are absorbed, distributed, metabolised, and eliminated in the body (pharmacokinetics).

(4) Synthesis of Drugs: The synthesis of APIs relies heavily on the principles of organic chemistry. These substances can be made in a consistent and repeatable fashion using chemical processes.

5. Quality Assurance: Pharmaceuticals rely on analytical chemistry to guarantee their potency, purity, and overall quality. Methods like high-performance liquid chromatography (HPLC) and mass spectrometry are used in quality control laboratories to ensure that pharmaceuticals are true to label.

The Responsibility and Effects of Chemistry on the Environment

The environment is profoundly affected by chemistry, which in turn influences human health. For the sake of the planet, we must practise environmentally responsible chemistry:

The State of the Air: To comprehend and control air pollution, chemistry is necessary. It aids in the detection and quantification of contaminants like smoke particles, VOCs, and greenhouse gases.

The State of the Water Supply In order to detect pollutants including heavy metals, herbicides, and pathogens, analytical chemistry is employed to keep an eye on the cleanliness of water supplies. This is essential for preventing water-related illnesses and maintaining clean drinking water supplies.

3. Renewable Sources of Energy Solar cells, wind turbines, and hydrogen fuel cells are just a few examples of sustainable energy sources that have benefited from chemistry's contributions. These innovations mitigate climate change by lessening the ecological footprint of the energy sector.

Managing Garbage 4 Waste treatment and recycling systems benefit from chemistry's input in their development. Decreased waste volume and lessened environmental damage are both possible thanks to chemical reactions and processes.

Chemistry and the Safety of Common Household Items.

Cleaning supplies, personal care items, and cosmetics are just some examples of everyday items whose composition and safety are influenced by chemistry:

First, the product must be formulated; formulas are created by chemists who strike a balance between efficacy and safety. They use the chemicals' properties to determine which additives, detergents, surfactants, and preservatives to use.

Second, all goods must pass stringent safety tests, such as those for skin and eye irritation, before they can be sold to the public.

Thirdly, sustainable chemistry strives to lessen the environmental toll of consumer goods by doing things like cutting back on potentially dangerous chemicals and developing more efficient packaging.

Chemistry and Commonplace Materials

Materials science is an interdisciplinary field that uses chemistry extensively to create novel materials for commonplace applications:

First, Textiles and Apparel: Synthetic fibres, such as polyester and nylon, are made via a chemical process. It is also used in stain-resistant and moisture-wicking fabric treatments.

Second, electronics: materials with specific electrical properties are essential to the operation of electronic gadgets like smartphones. The advancement of semiconductors, conductors, and insulators relies heavily on chemistry.

Third, polymers and plastics: Plastics and polymers, which are utilised in everything from food packaging to medical implants, have their origins in chemistry.

4. Materials for Buildings: Materials science and chemistry are the driving forces behind the creation of modern building materials including concrete additives and insulating materials.

Using Chemistry to Improve Healthcare and Diagnostics

Beyond drugs, chemistry plays an important role in healthcare. It plays an important role in medical imaging, diagnostics, and laboratory testing:

* 1. Diagnostic Radiology * Imaging methods that produce images of the inside of the body, such as MRI, X-rays, and CT scans, are all based on chemical principles.

Tests at a Clinical Laboratory Blood, urine, and other physiological fluids are analysed with chemical tests in clinical chemistry laboratories for diagnostic purposes. From diabetes to cancer, these tests are invaluable.

Biotechnology, No. 3 Molecular biology and chemistry are at the heart of biotechnology, which is used to create cutting-edge diagnostic instruments, gene therapies, and personalised medicine strategies.

Finally, the Pervasive Impact of Chemistry

Our daily life and health are profoundly influenced by chemistry. What we consume, what we take, how clean our surroundings are, and how secure the things we use are all influenced by this. The advancements in health care and environmental protection that we enjoy today are all thanks to chemistry's role as a catalyst for change. As our understanding of the world grows, it becomes an even more important part of daily life and a potent agent of change. The significance of chemistry in our lives highlights the need for ethical research and a dedication to applying scientific findings to improve human well-being.

Chapter 11.
Chemistry's Bright Future in Life Sciences

11.1- Current research and emerging fields where chemistry and biology intersect.

Current and Future Areas of Study at the Crossroads of Chemistry and Biology

Research at the interface between chemistry and biology is flourishing and producing exciting new findings and breakthroughs. Increased knowledge of biological systems and new treatments, diagnostics, and technology have resulted from this blending of academic fields. To demonstrate the revolutionary potential of this interdisciplinary strategy, we will delve into recent developments and promising new areas of study at the interface of chemistry and biology.

Chemical Biology: A Way to Bring People Together

The goal of the interdisciplinary study known as chemical biology is to apply the methods and concepts of chemistry to the study of biology in order to gain insight into and control over biological processes. The field is concerned with the creation of chemical probes and tiny molecules to investigate and control biological processes. Studies in chemical biology that are currently underway include:

Specific Drug Exploration: To selectively interact with their biological targets, such as proteins or nucleic acids, chemical biologists develop and synthesise tiny molecules. These chemicals are being studied as possible treatments for a wide variety of ailments, from cancer to neurological conditions.

Genomics of Chemical Substances: The goal of chemical genomics is to determine the roles that genes and proteins play in the body through the systematic testing of tiny molecules. This method is useful for locating disease-related pathways and prospective therapeutic targets.

- Chemical Probes: These are molecular tools created by scientists to better understand intricate biological processes. By targeting individual biomolecules within living cells, these probes make targeted manipulation and visualisation possible.

Molecular landscapes, uncovered through structural biology

It is the goal of the scientific discipline known as structural biology to discover the three-dimensional structures of biological macromolecules, and this is accomplished by employing methods from chemistry, biology, and physics. Some recent structural biology developments are:

Electron microscopy using liquid nitrogen (Cryo-EM) Because of the near-atomic resolution of cryo-EM, huge and complicated biomolecules like membrane proteins and ribosomes can finally be visualised by structural biologists. This method has provided fresh opportunities for finding novel medicines and creating new vaccines.

X-ray crystallography continues to be an effective method for elucidating the three-dimensional structures of crystalline biological macromolecules. The accuracy and productivity of this method have recently benefited from advancements in data collection and processing.

Insights into the structures and dynamics of biomolecules in solution can be gained using nuclear magnetic resonance (NMR) spectroscopy. The development of NMR has allowed it to be used on more elaborate systems.

Three, Synthetic Biology: The Art of Creating New Life

The goal of synthetic biology is to create biological systems with novel functions by combining biological and chemical principles. Current synthetic biology studies include:

Genetic Modification: CRISPR-Cas9 and other methods of precise genome editing have changed the face of genetic engineering. Researchers can now precisely alter the DNA of organisms, allowing for novel applications in gene therapy and biotechnology.

The Field of Bioengineering Bacteria that can create biofuels or enzymes for industrial processes are only two examples of the modified creatures made possible by synthetic biology. Both sustainable agriculture and bio-manufacturing can benefit from these genetically modified organisms.

Biology-Based Detection Devices To detect particular chemicals or environmental conditions, synthetic biologists create biological sensors. These sensors can be used for biosecurity, medical diagnosis, and monitoring the environment.

Chemical ecology has been called "the chemical language of nature."

The field of chemical ecology studies how chemicals affect living things and their surroundings. Chemical signalling in communication, defence, and resource acquisition is the focus of this study. Some of the most recent findings in chemical ecology are:

Hormones (Pheromones) In order to communicate with other members of their own species, animals employ pheromones, which are chemical messages. The chemical ecology field has shed light on the significance of pheromones in a variety of social and sexual contexts.

Chemical defences in plants: Secondary metabolites, such as alkaloids and terpenoids, are produced by plants to ward against herbivores and diseases. Sustainable farming methods can be developed by learning about these chemical defences.

Interactions Between Microorganisms: Chemical signalling is used by microbes to help them work together in complex communities. The potential for this area to aid in the creation of biofuels and probiotics is exciting.

The Brain's Chemical Codes (Chemical Neurobiology, Fifth Edition)

The study of chemical processes that underlie brain function and neurological illnesses is known as chemical neurobiology. Some of the most recent studies in this area include:

Signalling via Neurotransmitters Neurotransmitters and their receptors are chemical messengers whose chemistry must be deciphered if brain activity is to be understood. Neurotransmitters including dopamine and serotonin have been shown to have important roles in emotion regulation, cognition, and addiction.

The study of neuropharmacology. To combat neurological and mental diseases like depression, schizophrenia, and Alzheimer's, scientists are working on novel medications that specifically target neurotransmitter systems.

Neuroimaging, or - Functional magnetic resonance imaging (fMRI) and positron emission tomography (PET) are two examples of cutting-edge imaging technologies that give researchers a window into the brain's electrical activity and chemical processes in real time. These instruments are useful for detecting and tracking neurological conditions.

Chemical Ecology: The Language of Life's Chemistry

The field of chemical ecology studies how chemicals affect living things and their surroundings. Chemical signalling in communication, defence, and resource acquisition is the focus of this study. Some of the most recent findings in chemical ecology are:

Hormones (Pheromones) In order to communicate with other members of their own species, animals employ pheromones, which are chemical messages. The chemical ecology field has shed light on the significance of pheromones in a variety of social and sexual contexts.

Chemical defences in plants: Secondary metabolites, such as alkaloids and terpenoids, are produced by plants to ward against herbivores and diseases. Sustainable farming methods can be developed by learning about these chemical defences.

Interactions Between Microorganisms: Chemical signalling is used by microbes to help them work together in complex communities. The potential for this area to aid in the creation of biofuels and probiotics is exciting.

Chemistry of the Nervous System and the Brain:

The study of chemical processes that underlie brain function and neurological illnesses is known as chemical neurobiology. Some of the most recent studies in this area include:

Signalling via Neurotransmitters Neurotransmitters and their receptors are chemical messengers whose chemistry must be deciphered if brain activity is to be understood. Neurotransmitters including dopamine and serotonin have been shown to have important roles in emotion regulation, cognition, and addiction.

The study of neuropharmacology. To combat neurological and mental diseases like depression, schizophrenia, and Alzheimer's, scientists are working on novel medications that specifically target neurotransmitter systems.

Neuroimaging, or - Functional magnetic resonance imaging (fMRI) and positron emission tomography (PET) are two examples of cutting-edge imaging technologies that give researchers a window into the brain's electrical activity and chemical processes in real time. These instruments are useful for detecting and tracking neurological conditions.

Chemical Ecology: The Language of Life's Elements

The field of chemical ecology studies how chemicals affect living things and their surroundings. Chemical signalling in communication, defence, and resource acquisition is the focus of this study. Some of the most recent findings in chemical ecology are:

Hormones (Pheromones) In order to communicate with other members of their own species, animals employ pheromones, which are chemical messages. The chemical ecology field has shed light on the significance of pheromones in a variety of social and sexual contexts.

Chemical defences in plants: Secondary metabolites, such as alkaloids and terpenoids, are produced by plants to ward against herbivores and diseases. Sustainable farming methods can be developed by learning about these chemical defences.

Interactions Between Microorganisms:

Chemical signalling is used by microbes to help them work together in complex communities. The potential for this area to aid in the creation of biofuels and probiotics is exciting.

Chapter 9 Chemical Neurobiology: The Brain's Code

The study of chemical processes that underlie brain function and neurological illnesses is known as chemical neurobiology. Some of the most recent studies in this area include:

Signalling via Neurotransmitters Neurotransmitters and their receptors are chemical messengers whose chemistry must be deciphered if brain activity is to be understood. Neurotransmitters including dopamine and serotonin have been shown to have important roles in emotion regulation, cognition, and addiction.

The study of neuropharmacology. To combat neurological and mental diseases like depression, schizophrenia, and Alzheimer's, scientists are working on novel medications that specifically target neurotransmitter systems.

Neuroimaging, or - Functional magnetic resonance imaging (fMRI) and positron emission tomography (PET) are two examples of cutting-edge imaging technologies that give researchers a window into the brain's electrical activity and chemical processes in real time. These instruments are useful for detecting and tracking neurological conditions.

Chemical Ecology: The Language of Life.

The field of chemical ecology studies how chemicals affect living things and their surroundings. Chemical signalling in communication, defence, and resource acquisition is the focus of this study. Some of the most recent findings in chemical ecology are:

Hormones (Pheromones) In order to communicate with other members of their own species, animals employ pheromones, which are chemical messages. The chemical ecology field has shed light on the significance of pheromones in a variety of social and sexual contexts.

Chemical defences in plants: Secondary metabolites, such as alkaloids and terpenoids, are produced by plants to ward against herbivores and diseases. Sustainable farming methods can be developed by learning about these chemical defences.

Interactions Between Microorganisms: Chemical signalling is used by microbes to help them work together in complex communities. The potential for this area to aid in the creation of biofuels and probiotics is exciting.

11. Chemical Neurobiology: The Brain's Encoders

The study of chemical processes that underlie brain function and neurological illnesses is known as chemical neurobiology. Some of the most recent studies in this area include:

Signalling via Neurotransmitters Neurotransmitters and their receptors are chemical messengers whose chemistry must be deciphered if brain activity is to be understood. Neurotransmitters including dopamine and serotonin have been shown to have important roles in emotion regulation, cognition, and addiction.

The study of neuropharmacology. To combat neurological and mental diseases like depression, schizophrenia, and Alzheimer's, scientists are working on novel medications that specifically target neurotransmitter systems.

Neuroimaging, or - Functional magnetic resonance imaging (fMRI) and positron emission tomography (PET) are two examples of cutting-edge imaging technologies that give researchers a window into the brain's electrical activity and chemical processes in real time.

These instruments are useful for detecting and tracking neurological conditions.

The ever-changing landscape of chemical biology: a conclusion.

Innovations in medicine, biotechnology, and environmental study have all been fueled by the fertile ground where chemistry and biology meet. From finding tailored therapeutics for diseases to harnessing the potential of synthetic biology for sustainable solutions, current research trends and new fields within this transdisciplinary environment hold promise in tackling some of the most serious challenges facing science and society. New potential for discovery and revolutionary influence will emerge as the lines between chemistry and biology continue to blur as technology develops and our understanding grows. Where the mysteries of life and the possibilities for new discoveries meet is where the future of scientific research will be forged.

11.2- The potential for future breakthroughs and discoveries.

Future Opportunities for Innovation and Discovery

Discoveries, innovations, and new avenues of inquiry characterise the scientific method. At this moment in time, the prospects for revolutionary advances in many areas of science and technology are nothing short of astounding. In this adventure, we will travel to a world where scientific inquiry leads us to new boundaries and modifies our view of the cosmos, dreaming of a future where these possibilities become reality.

First, Astrophysics and Cosmology: Deciphering the Cosmic Enigmas

We stand on the cusp of significant discoveries in astrophysics and cosmology that will expand our knowledge of the cosmos and its history. Future developments in these areas could include:

Dark Energy and Dark Matter: There is still a lot we don't know about dark matter and dark energy, which together account for most of the mass-energy in the cosmos. Potentially game-changing insights into these mysterious phenomena could be gained from future investigations like the Large Synoptic Survey Telescope (LSST).

Exploration of Extrasolar Planets: Extraterrestrial life and the discovery of habitable exoplanets are at the cutting edge of astrophysics. There are tantalising hints regarding the habitability of distant exoplanets that could be detected and analysed with the help of future missions like the James Webb Space Telescope (JWST) and developments in spectroscopy.

The Effects of Gravitational Waves With the discovery of gravitational waves, a new era in astronomy has begun. The study of black hole mergers and the early universe, for example, will be made possible by current and future observatories like the Laser Interferometer Space Antenna (LISA).

Second, Breakthrough Medical Technology and Its Impact on Healthcare

Potentially game-changing advances in healthcare in the future hold great promise for improving diagnosis, treatment, and illness prevention.

The Promise of Precision Medicine Precision medicine will be the norm once genomes, proteomics, and individualised diagnostics reach critical mass. Individualised medicines that take into account a patient's unique genetic and molecular makeup have been shown to improve treatment outcomes while reducing unwanted side effects.

The Use of Immunotherapy: Cancer and autoimmune illnesses may be effectively treated with immunotherapy approaches like cancer vaccines and CAR-T cell treatments. Potentially improved therapy efficacy and a wider variety of treatable illnesses may result from future developments.

The Field of Neuroscience: The study of the brain is at the cutting edge of the field of medicine. Improvements in the treatment of neurological illnesses and the enhancement of cognitive capacities may result from advances in neuroimaging, neuromodulation, and brain-computer interfaces.

3 Green Technologies for Long-Term Energy and Environmental Security

The development of green technology is being propelled by the need for long-term energy security and environmental protection.

- Renewable Energy: Recent developments in solar cells, wind turbines, and energy storage systems are increasing the viability and practicality of this alternative energy option. The shift to a sustainable energy future may be hastened by recent advances in materials science and energy conversion.

Capturing and storing carbon dioxide: To slow global warming, we need to perfect carbon capture and storage technologies. Direct air capture and innovative materials for carbon capture are two examples of the kinds of cutting-edge methods that could completely change the landscape of this industry.

Remediating Polluted Areas using Biotechnology: The use of biotechnology to address environmental problems holds great promise. Bioremediation strategies and engineered microbes can work together to restore plastic-contaminated environments.

Quantum technologies: a catalyst for revolutionary change

When it comes to computing, communication, and cryptography, quantum technologies are set to usher in a new era.

Computing on the Quantum Level: Thanks to their enormous processing capacity, quantum computers may one day help us crack difficult issues in areas like encryption, drug discovery, and materials science. The time when quantum computers are more powerful than classical ones is rapidly approaching.

To communicate on a quantum level: By distributing keys in a quantum fashion, quantum communication provides extremely safe channels of communication. The development of quantum networks and the ideas behind the quantum internet hold the promise of secure global communication and unbreakable encryption.

Technology for Quantum Sensing and Imaging: Navigation, precise metrology, and the detection of small environmental changes are just a few of the many uses for quantum sensors including atomic clocks and quantum-enhanced imaging systems.

Intelligent Technologies, including AI and ML, as the Fifth Discipline.

Synergy between AI/ML and other scientific fields has the potential to propel game-changing developments.

Accelerating the development of new treatments and streamlining clinical trials is one of the many benefits of AI-powered drug discovery platforms. The pharmaceutical sector will undergo a radical transformation due to the rise of tools like virtual drug screening, predictive modelling, and precision medicine.

Revolutions in materials science are being driven by artificial intelligence. High-performance materials in fields as diverse as electronics and renewable energy can be designed with the help of algorithms that predict the features of innovative materials.

Modelling the Climate: Understanding and coping with climate change require climate models that have been advanced by artificial intelligence. Better climate projections, a deeper understanding of massive datasets, and the most effective ways to cut carbon emissions may all be achieved with the help of machine learning.

6. Exploration and Colonisation of Space: Broadening Humanity's Perspectives

The potential of space travel in the future is limitless:

- Settlement of Mars: Projects like SpaceX's Starship aspire to permanently settle humans on Mars. This effort has the potential to usher in a new era of space exploration.

Tourism in outer space: It's getting close to the point where commercial space travel can actually happen. Space tourism has the potential to boost both the economy and the scientific community as costs for space travel drop.

- Life From Outer Space: Missions to Jupiter's moon Europa and Saturn's moon Enceladus, both of which are believed to have subsurface oceans, are part of the ongoing hunt for extraterrestrial life. Even if it's just bacteria, finding life on another planet would be a huge step forward in science.

Responsibility in an Age of Ethics and Regulations

The importance of ethical considerations and regulatory frameworks grows as we explore new scientific regions.

The Ethics of AI There is a compelling need to ensure the ethical application of AI and machine learning.

Responsible AI development requires attention to bias, privacy, and openness.

Ethical concerns concerning designer babies and unexpected repercussions are being raised in light of recent developments in gene editing technologies such as CRISPR-Cas9. The world must act immediately to establish norms for ethical gene editing.

Governance of Space Peaceful and long-term space travel will require international agreements and rules for the control of space traffic, the responsible use of space resources, and protection of the space environment.

Insight into the Limitless Possibilities of Human Curiosity

Future discoveries and advancements are only as restricted as the insatiable human desire to learn about and explore the universe. Science and technology continue to move us towards a future that holds the promise of unprecedented knowledge, creativity, and development, from elucidating the secrets of the cosmos to harnessing the power of quantum physics and advancing the frontiers of medicine. To make sure that our discoveries and breakthroughs benefit all of mankind and maintain the fragile balance of our planet's ecosystems, we must keep a keen eye out as we venture into the unknown, guided by ethical standards and a dedication to responsible innovation. The future is a blank slate, ready to be filled with the ideas and hard work of generations yet to come.

Chapter 12.
Final Thoughts on an Ongoing Trip:

12.1- Summarizing the importance of chemistry in understanding life.

The Role of Chemistry in Our Understanding of Life

Chemistry, sometimes called the "central science," is essential to understanding the workings of the universe and of life itself. Connecting the simplest elements of matter to the complex processes that characterise living things, it is an essential link in the chain of explanation. Chemistry is the lens through which we obtain fundamental insights into the nature of life, from the composition of biomolecules to the kinetics of cellular interactions. In this investigation, we will discuss a brief summary of the vital function that chemistry plays in several branches of biology.

1. The Building Blocks of Life: The Molecules

Molecules, which consist of atoms bound together in particular patterns, are the fundamental units of all forms of life. The basic framework for understanding these compounds, which include:

Notes on DNA and RNA: Nucleotides are the building blocks of DNA and its complementary molecule, RNA. The study of genetics and molecular biology relies heavily on our knowledge of the chemical structure of nucleic acids and the rules of base pairing.

Note: Proteins Proteins serve as the biological "workhorses," catalysing chemical reactions (enzymes) and providing structural support, among other roles. DNA and peptide bond chemistry set the rules for the order of amino acids in proteins.

Foods High in Carbohydrates: Carbohydrates are important because they provide both fuel and structure. Understanding the intricacies of carbohydrates, from simple sugars to complex polysaccharides, is made possible by chemistry.

Fats and Oils Fats and phospholipids are two types of lipids that play crucial roles in cell membranes and the storage of energy. Their chemistry-controlled hydrophobic and hydrophilic characteristics underpin their biological functions.

2. The Language of Life as Expressed by Molecular Interactions

The language of chemistry helps us make sense of the complex molecular interactions that power all living things. Concepts central to this field include:

Bonds in Chemistry The construction of molecular structures relies heavily on covalent, ionic, and hydrogen bonding. These bonds are what give biological molecules their stability and responsiveness.

Water-Repellent and Water-Attractive Interactions Hydrophobic and hydrophilic characteristics are essential to the behaviour of molecules in liquids. The folding of proteins, the formation of cell membranes, and the intracellular transit of molecules all rely on this event to function properly.

Electrostatic interactions occur between charged molecules and ions. The behaviour of ions in cellular processes like nerve signalling and muscle contraction relies on an appreciation of these forces.

Third, Metabolism and Energy: How Chemistry Keeps Us Alive

A intricate network of chemical reactions powers the metabolic processes necessary for life. Some fundamentals of energy and metabolism are:

It's all about the thermodynamics. Thermodynamics controls the movement of energy within living organisms. grasp the spontaneity and direction of chemical processes within cells requires a grasp of concepts like entropy and free energy.

Enzymes, to wit: Enzymes speed up chemical reactions since they are biological catalysts. Their ability to catalyse reactions is dependent on their ordered three-dimensional structures, which are established by chemical laws.

Currency for Energy: The cellular currency of energy is adenosine triphosphate (ATP). Energy storage and release by ATP, which makes cellular work possible, is chemically explained.

Fourth, the chemistry of cellular processes

Chemical reactions within cells orchestrate the activities that characterise living things. The fundamentals of chemistry shed light on the following processes:

Signal Transduction in Cells: Hormones and neurotransmitters are just two examples of the chemical signalling molecules used in cellular communication. Processes including cell proliferation, differentiation, and responsiveness to external stimuli can be better understood when their chemical composition is known.

DNA Replicate, Transcribe, and Translate Chemistry governs many aspects of cell biology, including DNA replication, gene transcription into RNA, and protein translation. Chemical interactions are important to the precision and control of these processes.

Respiration at the Cellular Level: Cellular respiration, the process through which glucose is converted into ATP, is essential to life. The chemical reactions leading up to this energy output are laid bare by chemistry.

Unlocking Therapeutic Potential with Medicinal Chemistry

The goal of medicinal chemistry is to improve human health by using chemical principles in medication development to treat disease. Among the most important facets of medicinal chemistry are:

Drug Research and Development Proteins and enzymes are only two examples of the kind of biological targets that medicinal chemists seek for and work to build molecules to interact with. New medications start with these chemical molecules.

- Drug Formulation: The stability, solubility, and bioavailability of pharmaceuticals rely heavily on the chemistry that goes into generating drug formulations.

Dosage schedules and drug creation are directed by the study of the pharmacokinetics (a medication's absorption, distribution, metabolism, and excretion in the body).

Research Progress: Chemistry's Importance in the Development of Knowledge 6.

Chemistry is the engine that powers the scientific discoveries that increase our understanding of the living world. For example:

- Analytical Techniques Modern analytical chemistry techniques, such as mass spectrometry, nuclear magnetic resonance (NMR) spectroscopy, and X-ray crystallography, give researchers new insight into the structures and properties of biomolecules.

- Emerging Technologies: Chemistry is the backbone of cutting-edge innovations like CRISPR-Cas9 gene editing, which opens up groundbreaking opportunities for studying and modifying genetic data.

Working Together From Different Disciplines Innovative, cross-disciplinary research is encouraged when chemistry is combined with other scientific fields such as biology, physics, and computer science.

The Importance of Chemistry in Life's Overall Understanding

Chemistry emerges as the thread that stitches together the complexities of life itself in the magnificent tapestry of scientific discovery. As a result, we may now unlock the potential for discoveries that can revolutionise medicine, technology, and our understanding of the natural world, and interpret the molecular blueprints of life. Chemistry's role in our understanding of life is not only theoretical; it forms the basis for innumerable scientific and medical advances. The language of chemistry will continue to be an indispensible guide as we explore the depths of the universe, revealing new and exciting wonders of life on Earth and beyond.

12.2- Encouraging continued exploration of this fascinating field.

Promoting Further Research into an Interesting Area

For millennia, people have been fascinated by the ever-changing and mind-boggling world of science. It's a place where new questions are raised and old ones are searched for, where the limits of human understanding are continually being tested. There is a plethora of fascinating realms to investigate, from the minute realm of quantum physics to the vastness of cosmology, from the complexities of the brain to the secrets of the universe. It is crucial for the development of human knowledge and the improvement of society as a whole to encourage more research into this intriguing topic.

The possibility of revolutionary discoveries that can affect our future is a major argument in favour of maintaining a vigorous scientific research community. From the invention of electricity to the creation of vaccinations, scientific discoveries have changed human life forever. These advancements have not only raised the standard of living, but also prevented countless deaths. Antibiotics are just one example of how scientific progress has improved human health and longevity by treating diseases that were formerly fatal.

Innovations that have altered entire industries have also resulted from our growing awareness of the interplay between technology and the natural world. Examples include how the internet has changed the face of business and media. There have been many technological offshoots from space research, such as GPS and high-tech materials utilised in commonplace items. Encouraging scientists and researchers to explore the limits of our knowledge can result in revolutionary discoveries that could alter our world for the better.

Furthermore, scientific discovery is essential for addressing some of the world's most critical problems. One such pressing issue is climate change, which calls for in-depth knowledge of Earth's processes and the creation of novel approaches to addressing the problem. Scientists and researchers are at the vanguard of initiatives to lessen the effects of climate change through investigations into its root causes, simulations of its potential outcomes, and the creation of novel clean energy sources. More research needs to be done in this area before we can identify long-term answers that will keep future generations safe from harm.

Research in the scientific realm is vital not just for resolving issues in the natural world, but also for improving people's health. Research in the fields of biology, genetics, and medicine is constantly yielding novel insights and treatments for a wide variety of conditions. Biology and medicine are dynamic fields, with constant developments such as genome mapping and personalised medication. By encouraging research in these areas, we can help people all throughout the world get better care and live longer, healthier lives.

Additionally, scientific investigation encourages both economic development and creative thinking. Putting money into R&D helps advance technology, which opens up new fields and generates more employment possibilities. Countries that invest heavily in research and development are more likely to fare well in international trade. The United States, for instance, has led the way in the fields of technology and biotechnology due to its long history of scientific study and innovation.

Fostering intellectual curiosity and critical thinking, encouraging continued exploration in the scientific realm is a win-win. Science is more than just a body of knowledge; it is an ongoing endeavour of investigation and discovery. To improve our understanding of the natural world, scientists are continuously challenging current theories, creating new tests, and analysing collected data. Critical thinking, problem solving, and making decisions based on data are all fostered through this procedure. These abilities are useful in the scientific field, but they can also be applied in other fields like teaching, running a business, or making public policy.

Furthermore, scientific research has a collaborative and unified character that goes across national boundaries and cultural norms. Scientists from all across the world collaborate to find solutions to global problems, disseminate new information, and foster better relations between nations. The scientific community is a role model for international cooperation in an age when pandemics and climate change necessitate concerted action. These ties can be strengthened and a sense of common purpose can be fostered by encouraging nations to keep exploring this area.

However, there are a few obstacles that must be solved before we can hope to stimulate further investigation into the intriguing field of scientific inquiry. Funding is a significant obstacle. Spending heavily on resources like equipment, materials, and employees is commonplace in research initiatives. These endeavours typically have extended completion periods, making reliable finance crucial. The provision of funding for scientific research is largely dependent on government agencies, private foundations, and charitable organisations.

Scientific inquiry also needs to deal with moral dilemmas. Research must be undertaken ethically and in the best interests of society, and this is why strong ethical rules and monitoring are necessary as new technologies emerge in areas like genetics and artificial intelligence. Striking a balance between scientific advancement and ethical considerations is difficult but essential.

Greater public engagement and science communication is also required. In addition to the backing of scientists and governments, encouraging the research of scientific topics involves the participation of the general people. Efforts to explain complicated scientific topics to the public should be prioritised since scientific literacy and knowledge are crucial for making informed decisions. A lifelong appreciation for science can be sparked in young children by fostering their innate curiosity and awe.

In conclusion, supporting researchers in their quest to learn more about the world around them is crucial to solving the world's most pressing problems and improving people's quality of life. The results of scientific research have the ability to improve our understanding of the world and provide solutions to urgent problems like global warming and the healthcare crisis. It promotes international cooperation, helps young minds flourish, and helps the economy expand. However, a united effort by governments, organisations, scientists, and the public is necessary to overcome obstacles including funding, ethics, and science communication. We can unleash the full potential of human curiosity and keep probing the secrets of the universe if we invest in and prioritise scientific research.

www.ingramcontent.com/pod-product-compliance
Lightning Source LLC
LaVergne TN
LVHW020440070526
838199LV00063B/4797